# BEGINNERS GUIDE TO JAVA PROGRAMMING



Dr. Marrynal S Eastaff
Dr. V. Saravanan | Mrs. A. Gowri

SHANLAX
#SINCE1990
SHANLAX
PUBLICATIONS

| | | |
|---|---|---|
| Title | : | BEGINNERS GUIDE TO JAVA PROGRAMMING |
| Author's Name | : | Dr. Marrynal S Eastaff<br>Dr. V. Saravanan & Mrs. A. Gowri |
| Published by | : | Shanlax Publications,<br>Vasantha Nagar, Madurai - 625003,<br>Tamil Nadu India |
| Publisher's Address | : | 61, 66 T.P.K. Main Road,<br>Vasantha Nagar, Madurai - 625003,<br>Tamil Nadu, India |
| Printer's Details | : | Shanlax Press, 66 T.P.K. Main Road,<br>Vasantha Nagar, Madurai - 625003,<br>Tamil Nadu, India |
| Edition Details (I,II,III) | : | I |
| ISBN | : | 978-93-91373-84-9 |
| Month & Year | : | November, 2021 |
| Copyright @ | : | Dr. Marrynal S Eastaff<br>Dr. V. Saravanan & Mrs. A. Gowri |
| Pages | : | 212 |
| Price | : | ₹290./- |

# ACKNOWLEDGEMENT

I wish to express my gratitude to my Lord Almighty and my family for having helped me in compiling this Book. I also wish to express my gratitude to the Management, CEO and Principal, Hindusthan College of Arts and Science for their constant support and encouragement.

**Dr. Marrynal S Eastaff**
**Dr. V Saravanan**
**Mrs. Gowri A**

# CONTENTS

# CHAPTER 1
# Introduction to Object Oriented Programming

## 1 Introduction:

- Java is an object oriented language developed by Sun Microsystems.
- Around 1990 James Gosling, Bill Joy and others at Sun Microsystems began developing a language called Oak.
- Oak language was primarily developed to control microprocessors embedded in consumer items like PDA's for this language should be a platform independent, reliable and compact language.
- But however due to the sudden explosion of Internet and similar technologies, Sun Microsystems changed their project objectives and renamed the Oak language into Java.
- In 1994 the project team developed a web browser called "HotJava" capable of locating and running applets, a java program that can run on a web browser.

## 1.1 Object-Oriented Paradigm:

- The major objective of object-oriented approach is to eliminate some of the flaws encountered in the procedural approach
- OOP treats data as a critical element in the program development and does not allow it to flow freely around the system.
- It ties data more closely to the functions that operate on it and protects it from unintentional modification by other functions.

ISBN: 978-93-91373-84-9

- OOP allows us to decompose a problem into a number of entities called Objects and then build data and functions around these entities.
- The combination of data and methods make up an object.
- Some of the features of object-oriented paradigm are
  a. Emphasis is on data rather than procedure
  b. Programs are divided into what are known as Objects
  c. Data structures are designed such that they characterize the objects
  d. Methods that operate on the data of an object are tied together in the data structure
  e. Data is hidden and cannot be accessed by external functions
  f. Objects may communicate with each other through methods
  g. New data and methods can be easily added whenever necessary
  h. Follows bottom-up approach in program design
- Definition of Object-Oriented programming - Object-Oriented programming is an approach that provides a way of modularizing programs by creating partitioned memory areas for both data and functions that can be used as templates for creating copies of such modules on demand.

## 1.2 Basic Concepts of Object-Oriented Programming:

- Objects and Classes
- Data Abstraction and Encapsulation
- Inheritance
- Polymorphism

ISBN: 978-93-91373-84-9

- Dynamic Binding
- Message Communication

### 1.2.1 Objects and Classes

- Objects are the basic runtime entities in an object-oriented system.
- They may represent a person, a place, a bank account, a table of data or any item that the program may handle.
- Program objects should be chosen such that they match closely with the real-world objects.
- When a program is executed, the objects interact by sending messages to one another.
- For example, 'customer' and 'account' are two objects in a banking program, then the customer object may send a message to the account object requesting for the balance. Each object contains data and code to manipulate the data. It is sufficient to know the type of message accepted and the type of response returned by the objects.
- The entire set of data and code of an object can be made a user-defined data type using the concept of class.
- A class may be thought of as a 'data type' and an object as a variable of that data type.
- Once a class is defined, we can create any number of objects belonging to that class. A class is thus a collection of objects of similar type

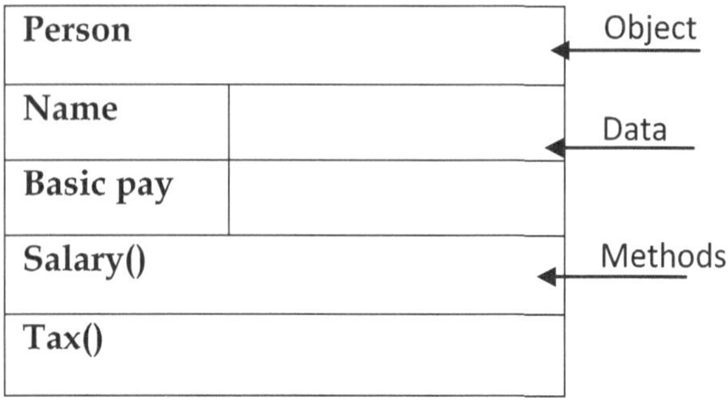


### 1.2.2 Data Abstraction and Encapsulation

- The wrapping up of data and methods into a single unit (called class) is known as encapsulation.
- The data is not accessible to the outside world and only those methods which are wrapped in the class, can access it.
- These methods provide the interface between the object's data and the program.
- This insulation of data from direct access by the program is called data hiding.
- Encapsulation is treated as 'black boxes' because the objects perform the specific task without knowledge of the internal implementation
- Abstraction refers to the act of representing essential features without including the background details or explanations.

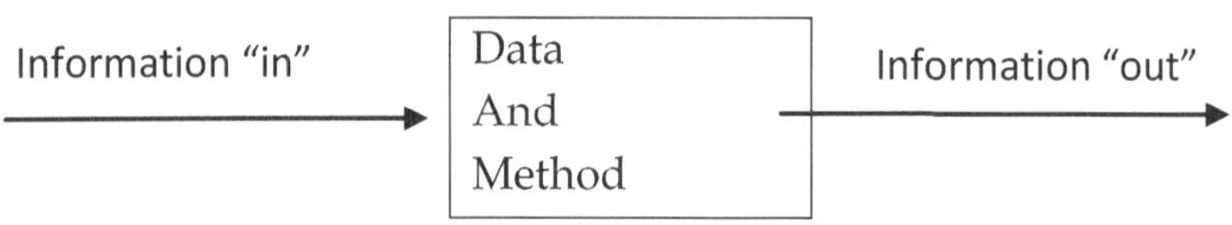


ISBN: 978-93-91373-84-9

### 1.2.3 Inheritance

- Inheritance is the process by which the objects of one class acquire the properties of objects of another class.
- Inheritance supports the concept of hierarchical classification.
- In OOP, the concept of inheritance provides the idea of reusability.
- This means that we can add additional features to an existing class without modifying it.
- The new class with have the combined features of both the classes.
- In Java, the derived class is known as 'subclass'

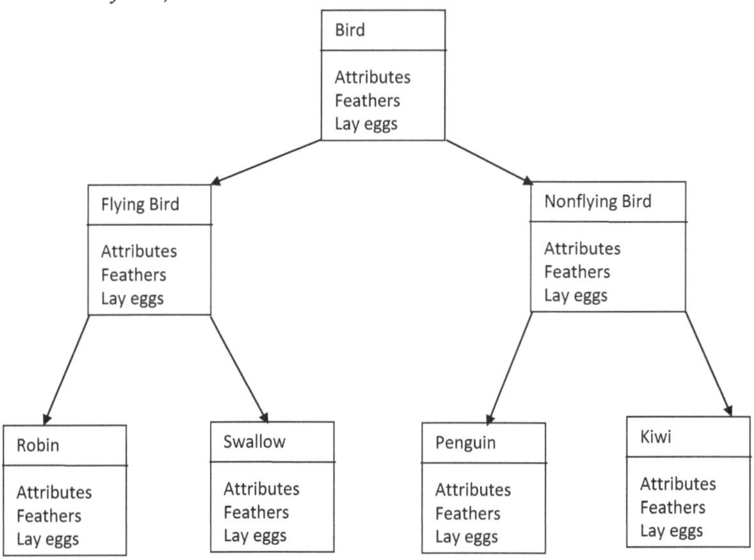


ISBN: 978-93-91373-84-9

### 1.2.4 Polymorphism

- Polymorphism is another important OOP concept. Polymorphism means the ability to take more than one form.
- For example, consider the operation of addition, for two numbers the operation will generate a sum. If the operands are strings, then the operation would produce a third string by concatenation.
- Polymorphism plays an important role in allowing objects having different internal structures to share the same external interface.
- Polymorphism is extensively used in implementing inheritance

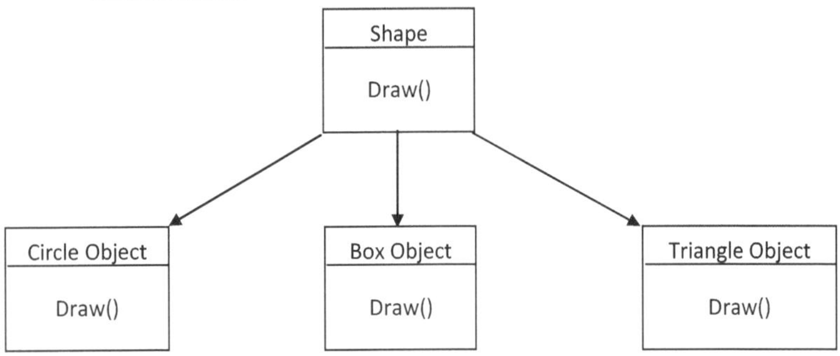


### 1.2.5 Dynamic Binding

- Java is a dynamic language. The linking of data and methods occurs only at run time. In java new classes can be loaded even when the program is on execution
- At run-time the code matching the object under current reference will be called

### 1.2.6 Message Communication

- An object-oriented program consists of a set of objects that communicate with each other. The process of

ISBN: 978-93-91373-84-9

programming in object-oriented language, involves three steps

  - Creating classes that define objects and their behaviour
  - Creating objects from class definitions
  - Establishing communication among objects

- Objects communicate with one another by sending and receiving information in the same way as people pass messages to one another
- A message for an object is a request for execution of a procedure, and therefore will invoke a method in the receiving object that generates the desired result.

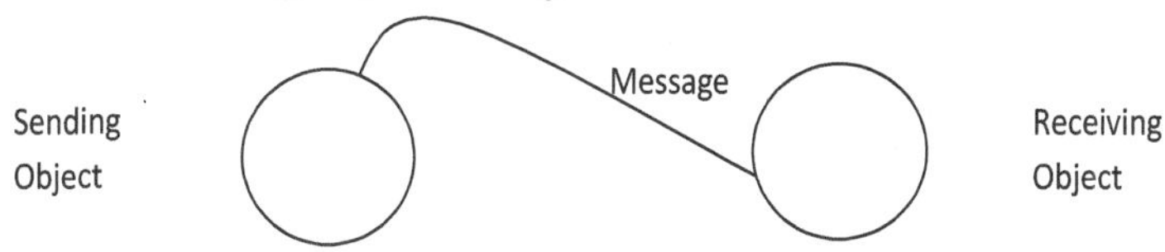


- Message passing involves specifying the name of the object, the name of the method (message) and the information to be sent.
- Example
  Employee.salary(name)

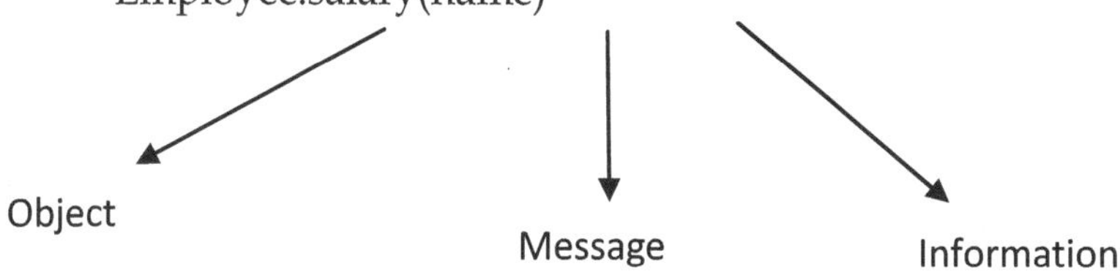


- Objects have a life cycle. They can be created and destroyed. Communication with an object is feasible as long as it is alive.

ISBN: 978-93-91373-84-9

## 1.3 Benefits of OOP

- OOP offers several benefits to both the programmer and the user. The principle advantages are
  - Through inheritance, we can eliminate redundant code and extend the use of existing classes
  - We can build programs from the standard working modules that communicate with one another, rather than starting the code from the scratch. This leads to saving of development time and higher productivity
  - The principle of data hiding helps the programmer to build secure programs.
  - It is easy to partition the work in a project based on objects
  - Object-oriented systems can be easily upgraded from small to large systems
  - Message passing techniques for communication between objects make the interface descriptions with external systems much simpler
  - Software complexity can be easily managed.

## 1.4 Applications of OOP

- The promising areas for application of OOP includes
  - Real-time systems
  - Simulation and modelling
  - Object-oriented databases
  - Hypertext, hypermedia
  - AI and expert system
  - Neural networks and parallel programming
  - Decision support and office automation systems
  - CIM/CA/CAD system

ISBN: 978-93-91373-84-9

## 1.5 Java Features

- Java is a high level language looks much similar to C, C++ but it offers many unique features that are not available in most of modern programming language.
- Java language has the following features.
- Compiled and Interpreted
- Platform Independence
- Object Oriented
- Robust
- Security
- Automatic Memory Management
- Dynamic Binding
- High Performance
- Multithreaded
- Some features in C/C++ eliminated

### 1.5.1 Compiled and Interpreted

- Java compiler converts the source code into intermediate byte code instructions.
- Java Interpreter converts these bytes codes into machine executable statements.
- Thus we can say java is both compiled and interpreted language. The two steps of compilation and interpretation allow for extensive code checking and improved security

### 1.5.2 Platform Independence

- Java programs are capable of running on all platforms hence java programs possess Write-Once-Run-Anywhere feature

ISBN: 978-93-91373-84-9

### 1.5.3 Object Oriented

- Java is purely an object oriented language. In Java no coding outside of class definitions is allowed, including the main() function.
- Java supports an extensive class library available in the core language packages.

### 1.5.4 Robust

- Java is a robust language. It provides the following facilities to ensure reliable code
    - Built in Exception Handling
    - Strong type checking both at compile and runtime.
    - Local variables must be initialized

### 1.5.5 Security

- Java is a highly secured language. Java programs will run only within the Java Virtual Machine (JVM) sandbox thus threats from malicious programs is minimized.
- Java does not support pointers hence direct memory access is not allowed.
- Java supports a built in security manager that determines what resources a class can access such as reading and writing to the local disk.

### 1.5.6 Automatic Memory Management

- Java supports automatic memory management. Java Virtual machine (JVM) performs garbage collection automatically to claim the memory occupied by unused objects.

### 1.5.7 Dynamic Binding

- Java is a dynamic language. The linking of data and methods occurs only at run time. In java new classes can be loaded even when the program is on execution.

ISBN: 978-93-91373-84-9

### 1.5.8 High Performance

- Interpretation of bytecodes slowed performance in early versions, but advanced virtual machines with adaptive and just-in-time compilation and other techniques now typically provide improved performance.

### 1.5.9 Multithreaded

- In java it possible to write programs to handle multiple tasks simultaneously. To write such programs you have to create light weight processes called "Threads" in java.
- The threads thus created can be executed simultaneously. Multithreaded programs are useful when you write programs for networking and multimedia.

### 1.5.10 Some Features in C/C++ Eliminated

- Some features in C/C++ are eliminated in java. They are listed below.
  - No Pointers
  - No Preprocessor Directives
  - No Multiple Inheritance in Classes
  - No template Classes

## 1.6 How Java Differs from C and C++

### 1.6.1 Java and C

- Java is a lot like C but the major difference between Java and C is that Java is an object-oriented language and has the mechanism to define class and objects
- The major difference is as follows

- Java does not include the C unique statement keywords sizeof and typedef
- Java does not contain the data type struct and union
- Java does not define the type modifiers auto, extern, register, signed and unsigned
- Java does not support an explicit pointer type.
- Java does not have a pre-processor and therefore we cannot use #define, #include and #ifdef statements
- Java requires that the functions with no arguments must be declared with empty parenthesis and not with the void keyword
- Java adds new operators such as instance of and >>>
- Java adds labelled break and continue statements
- Java adds many features required for object-oriented programming

### 1.6.2 Java and C++

- The major difference is as follows
    - Java does not support operator overloading
    - Java does not have template classes as in C++
    - Java does not support multiple inheritance of classes. This is accomplished using a new feature called "interface".
    - Java does not support global variables. Every variable and method is declared within a class and forms part of that class
    - Java does not use pointers
    - Java has replaced the destructor function with a finalize() function
    - There are no header files in java

ISBN: 978-93-91373-84-9

## 1.7 Java and Internet

- Java is strongly associated with the Internet because of the fact that the first application program written in Java was HotJava, a Web browser to run applets on Internet.
- Internet users can use Java to create applet programs and run them locally using a "Java-enabled browser".
- The Java enabled browser can also download an applet located on an remote computer and run it on their local computer.

## 1.8 Java and World Wide Web

- Java communicates with a Web page through a special tag called <APPLET>. The following figure shows the communication steps

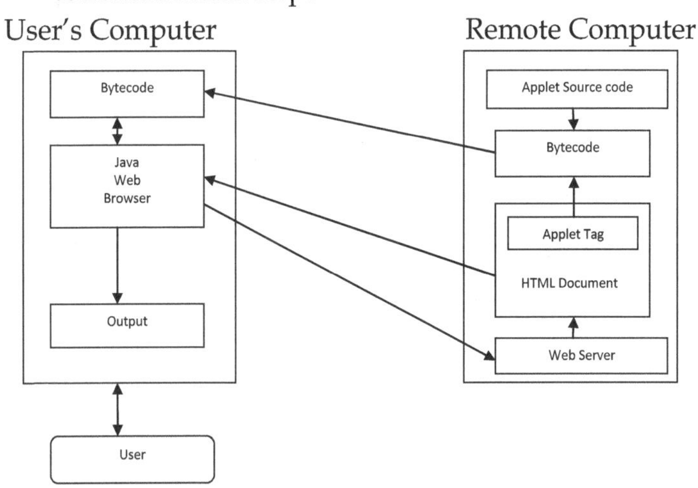


- The user sends a request for an HTML document to the remote computer's Web server. The web server is a program that accepts a request, processes the request and sends the required document.

- The HTML document is returned to the user's browser. The document contains the APPLET tag, which identifies the applet
- The corresponding applet bytecode is transferred to the user's computer. This bytecode had been previously created by the Java compiler using the Java source code file or that applet.
- The Java-enabled browser on the user's computer interprets the bytecodes and provides the output.
- The user may have further interaction with the applet but no further downloading from the web server. This is because the bytecode contains all the information necessary to interpret the applet.

## 1.9 Java Environment

- The Java environment includes a large number of development tools and hundreds of classes and methods
- The development tools are part of the system known as **Java Development Kit(JDK)**
- The classes and methods are part of the **Java Standard Library (JSL)** also known as **Application Programming Interface (API)**

### 1.9.1 Java Development Kit (JDK)

- This kit comes with a collection of tools that are used for developing and running Java programs. They include

ISBN: 978-93-91373-84-9

| Appletviewer | for viewing Java applets. Enables us to run Java applets (without actually using a Java-compatible browser) |
|---|---|
| javac | The java compiler, which translates Java source code to bytecode files that the interpreter can understand |
| java | Java Interpreter, which runs applets and applications by reading and interpreting bytecode files |
| javadoc | Creates HTML format documentation from Java source code files |
| javah | Produces header files for use with native methods |
| javap | Java disassemble, which enables us to convert bytecode files into a program description |
| jdb | Java debugger, which helps us to find errors in our program |

- The way these tools are applied to build and run application programs is illustrated below.
- To create a Java program we need to create a source code file using the text editor.
- The source code is then compiled using the Java compiler javac and executed using the Java interpreter java.
- The java debugger jdb is used to find errors if any, in the source doe.
- A compiler Java program can be converted into a source doe with the help of Java dissembler javap.

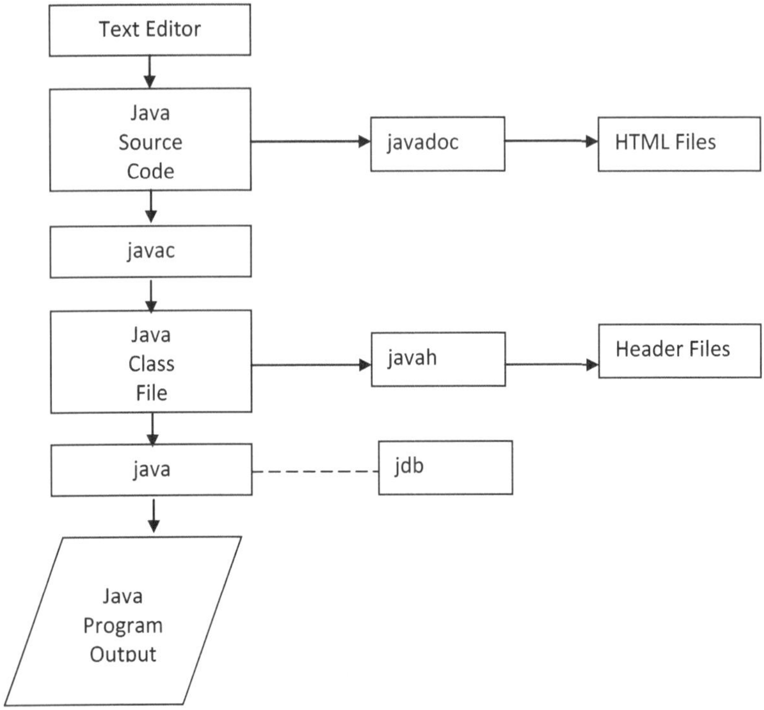


## 1.9.2 Application Programming Interface

- The Java Standard Library (or API) includes hundreds of classes and methods grouped into several functional packages. Most commonly used packages are
  - **Language Support Package -** A collection of classes and methods required for implementing basic features of Java
  - **Utilities Package-** A collection of classes to provide utility functions such as date and time functions
  - **Input/Ouput Package-**A collection of classes required for input/output manipulation
  - **Networking Package -**A collection of classes for communicating with other computers via Internet

ISBN: 978-93-91373-84-9

- **AWT Package-**The Abstract Window Tool Kit package contains classes that implements platform-independent graphical user interface
- **Applet Package-**This includes a set of classes that allows us to create Java applets

## 1.10 Overview of Java Language

- Java is a general-purpose, object-oriented programming language. We can develop two types of Java programs
  - Stand-alone applications
  - Web applets

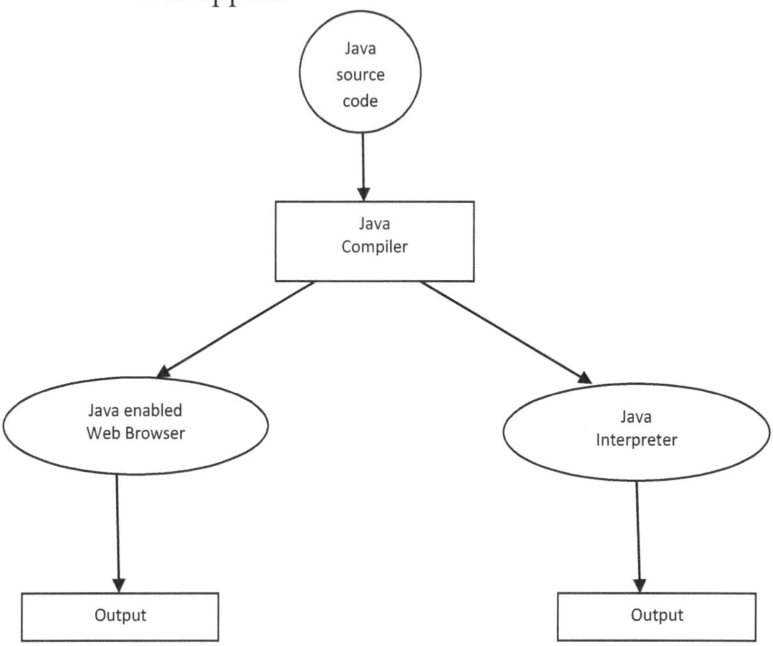


ISBN: 978-93-91373-84-9

- Stand-alone applications are programs written in Java to carry out certain task on a stand-alone local computer. Java can be used to develop programs for all kinds of applications
- Executing a stand-alone Java program involves two steps
  - Compiling source code into bytecode using Java compiler
  - Executing the bytecode program using java interpreter
- Applets are small Java programs developed for Internet applications. An applet located on a distant computer (Server) can be downloaded via Internet and executed on local computer (Client) using a Java-capable browser.
- Since applets are embedded in an HTML document and run inside a Web page, creating and running applets are more complex than creating an application.

### 1.10.1 Simple Java Program

- Consider the Java program to print a line of text

```
class Sample
{
        public static void main(String args[])
        {
                System.out.println("Java is better than C++");
        }
}
```

ISBN: 978-93-91373-84-9

#### 1.10.1.1 *Class Declaration*

- Java is a true object-oriented language and therefore, everything must be placed inside a class.
- Class is a keyword and declares that a new class definition follows. Sample is a Java identifier that specifies the name of the class to be defined.

#### 1.10.1.2 *Opening Braces*

- Every class definition in Java begins with an opening brace "{" and ends with a matching closing brace "}"
- In java the class is not terminated with a semicolon.

#### 1.10.1.3 *The Main Line*

- The following line defines a method named main
  - Public static void main(String args[])
- Every Java application program must include main() method. This is the starting point for the interpreter to begin the execution of the program.
- Java application can have any number of classes but only one of them must include the main method to initiate the execution
- The following keywords are used in the main method
  - Public -the keyword public is an access specifier that declares the main method as unprotected and therefore making it accessible to all other classes.
  - Static-the keyword static declares this method as one that belongs to the entire class and not a part of any objects of the class. The main must always be declared as static since the interpreter uses this method before any objects are created.
  - Void-the type modifier void states that the main method does not return any value

ISBN: 978-93-91373-84-9

#### *1.10.1.4 The Output Line*

- Since Java is a true object-oriented language, every method must be part of an object.
- The println method is a member of the out object, which is static data member of the System class.
- This line prints the string
    - Java is better than C++
- The method println always appends a newline character to the end of the string.
- Every Java statement must end with a semicolon.

### 1.11 Java Program Structure

- A Java program may contain many classes of which only one class defines a main method.
- Classes contain data members and methods that operate on the data members of the class
- Methods may contain data type declarations and executable statements.
- A Java program may contain one or more sections

| Document Section |
|---|
| Package Statement |
| Import statements |
| Interface Statements |
| Class Definitions |
| Main Method Class<br>{<br>Main Method Definition<br>} |

### 1.11.1 Document Section

- This section contains a set of comment lines giving the name of the program, the author and other details, which the programmer would like to refer to at a later stage.
- Comments inform the programmer what the program is for and what it does
- There are three styles of comments in Java
  - // - single line comment
  - /*........*/ - Multiline comment
  - /** ...*/ - documentation Comment
- This is used for generating documentation automatically

### 1.11.2 Package Statement

- The first statement allowed in a Java file is a package statement. This statement declares a package name and informs the compiler that the classes defined here belong to this package.
- Example
  - package student;

### 1.11.3 Import Statements

- The next thing after a package statement may be a number of import statements.
- This is similar to the #include statement in C
- Example
  - Import student.test;
- This statement instructs the interpreter to load the test class contained in the package student.

ISBN: 978-93-91373-84-9

### 1.11.4 Interface Statements

- An interface is like a class but includes a group of method declarations.
- This is also an optional section and is used only when we wish to implement the multiple inheritance features in the program.

### 1.11.5 Class Definitions

- A Java program may contain multiple class definitions.
- Classes are the primary and essential elements of a Java program. These classes are used to map the objects of real-world problems.

### 1.11.6 Main Method Class

- Since every Java stand-alone program requires a main method as its starting point, this class is the essential part of a Java program
- A simple Java program may contain only this part.
- The main method creates objects of various classes and establishes communications between them.
- On reaching the end of main, the program terminates and the control passes back to the operating system.

## 1.12 Java Tokens

- A Java program is basically a collection of classes. A class is defined by a set of declaration statements and methods containing executable statements.
- Most statements contain expressions which describe the actions carried out on data.
- Smallest individual units in a program are known as tokens. The compiler recognizes them for building up expressions and statements.

ISBN: 978-93-91373-84-9

- Java program is a collection of tokens, comments and white spaces. Java  language include five types of tokens. They are
  - Reserved Keywords
  - Identifiers
  - Literals
  - Operators
  - Separators

## 1.13 Java Character Set

- The smallest units of Java language are the characters used to write Java tokens. These characters are defined by the Unicode character set, an emerging standard that tries to create characters for a large number of scripts worldwide.
- The Unicode is a 16 bit character coding system and currently supports more than 34,000 defined characters derived from 24 languages.

### 1.13.1 Reserved Keywords

- Keywords are essential part of a language definition. Java language has reserved 50 words as keywords.
- Since keywords have specific meaning in Java, we cannot use them as name for variables, classes and methods.
- All keywords are to be written in lower-case letters.
- Example
  - Byteclassforgotodoetc

### 1.13.2 Identifiers

- Identifiers are programmer-designed tokens. They are used for naming classes, methods, variables, objects, labels, packages and interfaces in a program.
- Java identifiers follow the following rule
    - They can have alphabets, digits and the underscore and dollar sign characters
    - They must not begin with a digit
    - Uppercase and lowercase letters are distinct
    - They can be of any length
- Identifiers must be meaningful, short enough to be quickly and easily typed and long enough to be descriptive and easily read.

### 1.13.3 Literals

- Literals in Java are a sequence of characters (digits, letters, and other characters) that represent constant values to be stored in variables.
- Java language specifies five major types of literals. They are
    - Integer literals
    - Floating_point literals
    - Character literals
    - String literals
    - Boolean literals
- The type describes how the values behave and how they are stored.

### 1.13.4 Operators

- An operator is a symbol that takes one or more arguments and operates on them to produce a result.

ISBN: 978-93-91373-84-9

### 1.13.5 Separators

- Separators are symbols used to indicate where groups of code are divided and arranged. They basically define the shape and function of our code.
- The various separators are

  - Parentheses ( ) - used to enclose parameters in method definition and invocation. Also used for defining precedence in expressions etc.
  - Baraces { } - used to define a block of code for classes, methods and local scopes
  - Brackets [ ] - used to declare array types and for dereferencing array values
  - Semicolon ; - used to separate statements
  - Comma , - used to separate consecutive identifiers in a variable declaration
  - Period . - used to separate package names from sub-packages and classes

ISBN: 978-93-91373-84-9

## 1.14 Java Statements

- The statements in Java are like sentences in natural languages. A statement is an executable combination of tokens ending with a semicolon.
- Statements are usually executed in sequence in the order in which they appear.
- There are several types of statements

| Statement | Description |
|---|---|
| Empty statement | These do nothing and are used during program development as a place holder |
| Labelled Statement | Any statement may begin with a label. Labels in Java are used as the arguments of Jump statements |
| Expression Statement | Java has seven types of Expression statements<br>Assignment, Pre-Increment, Pre-Decrement, Post-Increment, Post-Decrement, Method Call and Allocation Expression |
| Selection Statement | These select one of several control flows. There are three types of selection statements<br>If, if – else, and switch |
| Iteration Statement | These specify how and when looping will take place. There are three types of iteration statements, while, do and for |
| Jump Statement | Jump statement pass control to the beginning or end of the current block. The four types of jump statement are break, continue, return and throw |
| Synchronization | These are used for handling issues with |

ISBN: 978-93-91373-84-9

| Statement | multithreading |
|---|---|
| Guarding Statement | Guarding statements are used for safe handling of code that may cause exceptions. These statements use the keywords try, catch and finally |

## 1.15 Implementing a Java Program

- Implementing a java program involves a series of steps. They are
  - Creating the program
  - Compiling the program
  - Running the program

### 1.15.1 Creating the Program

- We can create a program using a text editor. Save the file with the extension as .java.
- The file name should be the class name. This file is called the source file.
- Example

```
class test
{
        public static void main(String a[])
        {
                System.out.println("Welcome to Java
                Programming");
        }
}
```

- Save the file as test.java

### 1.15.2 Compiling the Program

- To compile the program, we must run the Java Compiler javac, with the name of the source file on the command line

ISBN: 978-93-91373-84-9 

- Example: javac test.java
- If the program is error free, the javac compiler creates a file called test.class containing the bytecodes of the program. The compiler automatically names the bytecode files as
  <classname>.class

### 1.15.3 Running the Program

- We need the Java Interpreter to run a stand-alone program. At the command prompt, type
  - java test
- The interpreter looks for the main method in the program and begins execution from there

### 1.15.4 Machine Neutral

- The compiler converts the source code files into bytecode files. These codes are machine-independent and therefore can be run on any machine.
- That is a program compiled on an IBM machine will run on a Macintosh machine
- Java interpreter reads the bytecode files and translates them into machine code for the specific machine on which the Java program is running.

## 1.16 Java Virtual Machine(JVM)

- All language compilers translate source code into machine code for a specific computer
- The Java compiler produces an intermediate code known as bytecode for a machine that does not exist.
- This machine is called the Java Virtual Machine and it exists only inside the computer memory.
- The process of compiling the Java program into bytecode is referred to as virtual machine code.

ISBN: 978-93-91373-84-9

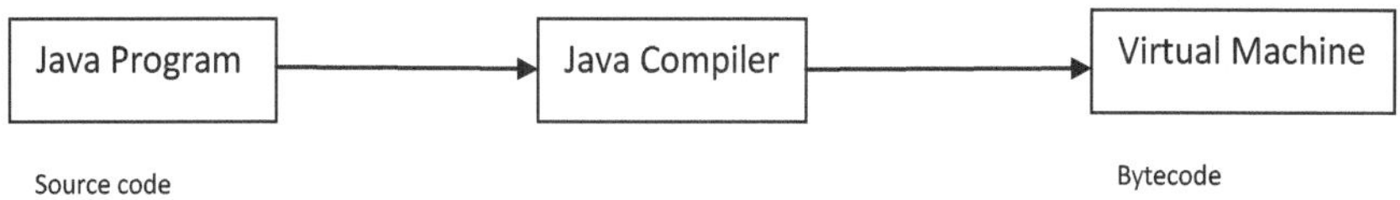


- The virtual machine code is not machine specific. The machine specific code is generated by the Java Interpreter by acting as an intermediate between the virtual machine and the real machine.

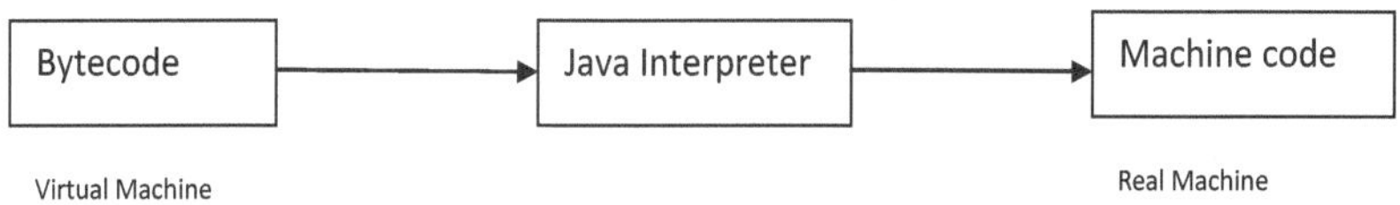


- Layers of interactions for Java Programs

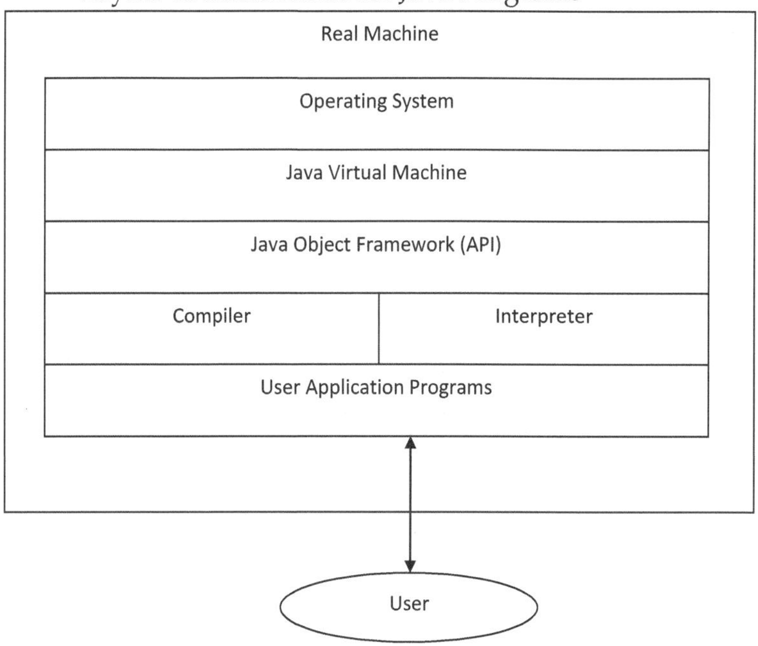


ISBN: 978-93-91373-84-9

# CHAPTER - 2
# Data Types, Operators and Expressions

## 2 Introduction

- A programming language is designed to process certain kinds of data and to provide useful output known as information.
- The task of processing data is accomplished by executing a sequence of instructions called a program.
- These instructions are formed using certain symbols and words according to some rigid rules known as syntax rules (or grammar)

## 2.1 Constants

- Constants are referred to fixed values that do not change during the execution of a program.

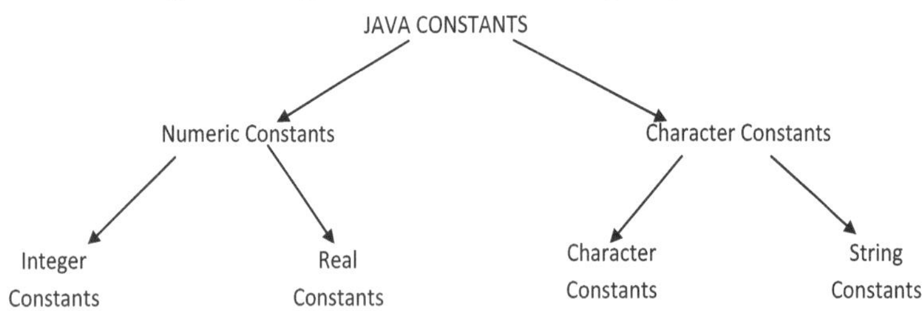


- Java supports several types of constants
  - Integer constants
  - Real constants
  - Single character constants
  - String constants
  - Backslash character constants

ISBN: 978-93-91373-84-9

### 2.1.1 Integer constants

- An integer constant refers to a sequence of digits. There are three types of integers namely
  - Decimal integer
  - Octal integer
  - Hexadecimal integer
- Decimal integers consist of a set of digits 0 through 9, preceded by an optional minus sign. Valid decimal integers are
  - 123, -321, 0
- Embedded spaces, commas and non-digit characters are not allowed between digits
- An octal integer consists of any combination of digits from the set 0 through 7, with a leading 0. Example
  - 037, 051, 0551
- A sequence of digits preceded by 0x or 0X is considered as hexadecimal integer. It includes 0 – 9, A-F. Example
  - 0x2, 0xbcd

### 2.1.2 Real Constants

- Quantities that vary continuously are represented using real constants.
- These quantities are represented by numbers containing fractional parts like 17.56
- Such numbers are called as real or floating point constants.
- These numbers are represented using a decimal notation, having a whole number followed by a decimal part and the fractional part.
- Example
  - 23.34, .45,

ISBN: 978-93-91373-84-9

- A real number may also be expressed in exponential or scientific notation. Example
  - 2.143e2
- The general form
  - Mantissa e exponent

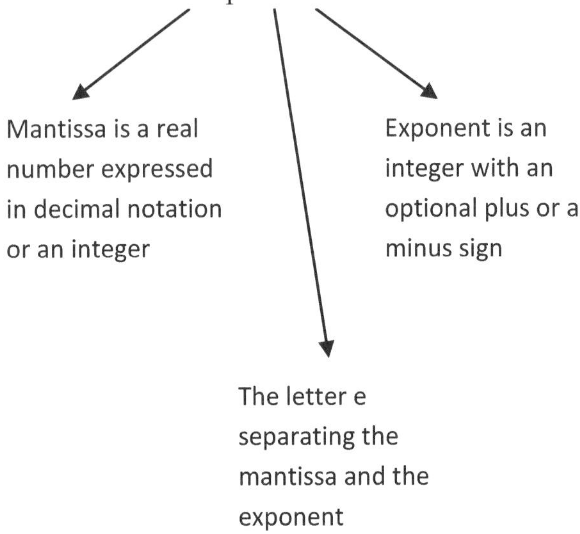


- A floating point constant has four parts
  - Whole number
  - A decimal point
  - A fractional part
  - An exponent

### 2.1.3 Single Character Constants

- A single character constant contains a single character enclosed within a pair of single quote marks. Example
  - '5', 'a', ';'

ISBN: 978-93-91373-84-9

### 2.1.4 String Constants

- A string constant is a sequence of characters enclosed between double quotes. The characters may be alphabets, digits, special characters and blank spaces.
  - Example: “hello”, “1997”,“welcome”

### 2.1.5 Backslash Character Constant

- Java supports some special backslash character constants that are used in output methods.
- These character combinations are known as escape sequences
- Example

| Constant | Meaning |
|---|---|
| ‘\b’ | Backspace |
| ‘\f’ | Form feed |
| ‘\n’ | New line |
| ‘\r’ | Carriage return |
| ‘\t’ | Horizontal tab |
| ‘\’’ | Single quote |
| ‘\”’ | Double quote |
| ‘\\’ | backslash |

## 2.2 Variables

- A variable is an identifier that denotes a storage location used to store a data value.
- A variable may take different values at different times during the execution of the program.
- A variable name can be chosen by the programmer in a meaningful way so as to reflect what it represents.
- Example: Average, height, a, weight

ISBN: 978-93-91373-84-9

### 2.2.1 Rules for writing variable names

- Variable name consists of alphabets, digits, the underscore and dollar characters
- They must not begin with a digit
- Uppercase and lowercase characters are distinct.
- It should not be a keyword
- White space is not allowed
- Variable names can be of any length

## 2.3 Data Types

- Every variable in Java has a data type. Data types specify the size and type of values that can be stored.
- Java language is rich in its data types.

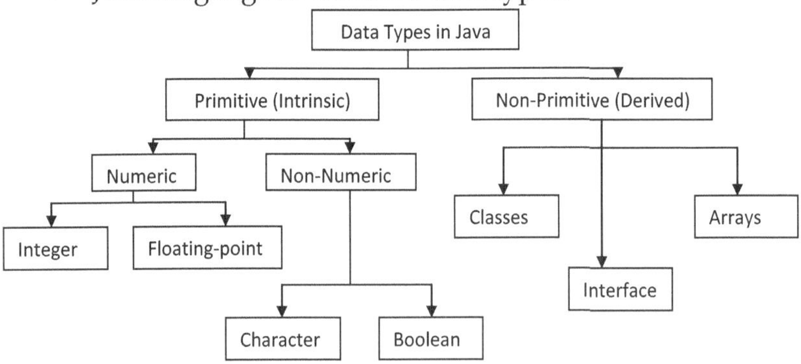


### 2.3.1 Integer types:

- Integer types can hold whole numbers such as 123, -34. The size of the values that can be stored depends on the integer data type we choose.
- Java supports 4 types of integers, they are byte short, int and long.
- Java does not support the concept of unsigned types and therefore all java values are signed, meaning they can be +ve or –ve

ISBN: 978-93-91373-84-9

| Type | Size | Range |
|---|---|---|
| byte | 1 byte | -128 to 127 |
| short | 2 bytes | -215 to 215-1 |
| int | 4 bytes | -231 to 231-1 |
| long | 8 bytes | -263 to 263-1 |

### 2.3.1.1 *byte:*

- The smallest integer type is byte. This is a signed 8-bit type that has a range from -128 to 127.
- Variables of type 'byte' are especially useful when you're working with a stream of data from a n/w or file.
- byte variables are declared by use of 'byte' keyword.
- eg: byte b, c;

### 2.3.1.2 *short:*

- short is signed 16 bit type. It has range from -215 to 215-1.
- The short variables are declared by use of short keyword.
- eg: short s;

### 2.3.1.3 *int:*

- The most commonly used integer type is int. It is a signed 32 bit type that a range from -231 to 231-1.
- In addition to other uses, variables of type int are commonly employed to control loop and to index arrays.
- 'int' type is most versatile & efficient type. int variables are declared by use of 'int' keyword.
- eg: int a;

ISBN: 978-93-91373-84-9

### 2.3.1.4 *long:*

- long is a signed 64 bit type and is useful for those occasions where an int type is not large enough to hold the desired values.
- The range of long is quite large. It has range from -263 to 263-1. long variables are declared by use of 'long' keyword
- eg: long a;

## 2.4 Floating point types:

- Floating point numbers are used when evaluating expressions that require fractional precision. There are two kinds of floating point types, float and double. Which represent single and double precision numbers respectively.

| Type | Size | Range |
|---|---|---|
| float | 4 bytes | 1.4e-045 to 3.4e+038 |
| short | 8 bytes | 4.9e-324 to 1.8e+308 |

### 2.4.1 float:

- The type float specifies a single precision value that uses 32 bits of storage. Single precision is faster on same processors and takes half as much space as double precision.
- float variables are declared by use of 'float' keyword.
- eg: float a=1.234;

### 2.4.2 double:

- double precision, as denoted by the double keyword, uses 64 bits to store a value.

ISBN: 978-93-91373-84-9

- All transcendental math functions such as sin(), cos() and sqrt() return double values. double variables are declared by use of 'double' keyword.
- eg: double d;

## 2.5 char:

- In java, the datatype used to store characters is char. Java uses Unicode to represent characters.
- Unicode defines a fully international character set that can represent all of the characters found in all human languages.
- Thus in java char is a 16bit type. The range of char is 0 to 65536.
- There are no negative chars. char variables are declared by use of char keyword.
- eg: char c;

## 2.6 boolean:

- Java has a simple type called boolean, for logical values. It can have only one of two possible values, true or false. This is the type returned by all relational operators such as a<b.
- Boolean is also the type required by conditional expressions that govern the control statements such as 'if' and 'for'. boolean variables are declared by use of 'boolean' keyword.
- eg: boolean b; b=false;

## 2.7 Declaration of Variables

- In Java, variables are the names of storage locations. Declarations does three things
  - It tells the compiler what the variable name is

ISBN: 978-93-91373-84-9

  - It specifies what type of data the variable will hold
  - The place of declaration decides the scope of the variable.
- The variable must be declared before it is used in the program
- A variable can be used to store a value of any data type.
- The general syntax of declaration of variables is

  type variable1, variable2, .... variable;
- Variables are separated by commas and the declaration statement must end with a semicolon.
- Example
  - int count;
  - float a,b,c;

## 2.8 Giving Values to Variables

- A variable must be given a value after it is declared and before it is used in an expression. This is achieved in two ways
  - By using an assignment statement
  - By using a read statement

## 2.9 Assignment Statement

- Simple method of giving value to a variable is throught the assignment statement
- Syntax

  variablename = value;
- Example
  - A=0;
  - Name="raja";
- It is also possible to assign values to variables at the time of its declarations

ISBN: 978-93-91373-84-9

- Syntax

  type variablename = value;
- Example
  - int a=0;
  - float b=3.27;
- The process of giving initial values to variables is known as initialization. The ones that are not initialized are automatically set to zero.

## 2.10 Read Statement

- We can also give values to variables interactively through the keyboard using the readLine() method

- Example

```
import java.io.DataInputStream;
class ex1
{
        public static void main(String args[])
        {
            DataInputStream in= new
        DataInputStream(System.in);
            int n=0;
            float f=0.0f;
            try
            {
                    System.out.println("Enter a Integer: ");
                    n=Integer.parseInt(in.readLine());
                    System.out.println("Enter  a float: ");

            f=Float.valueOf(in.readLine()).floatValue();
            }
            catch(Exception e)
            {       }
```

ISBN: 978-93-91373-84-9

```
            System.out.println("N= "+n);
            System.out.println("F= "+f);
        }
    }
```

- The readLine() method reads the input from the keyboard as a string which is then converted to the corresponding data type using data type wrapper classes.

## 2.11 Operators and Expressions

- An operator is a symbol that tells the computer to perform certain mathematical or logical manipulations
- Operators are used in programs to manipulate data and variables
- Java operators can be classified into a number of related categories
  - Arithmetic operators
  - Relational operators
  - Logical operators
  - Assignment operators
  - Increment and decrement operator
  - Conditional operator
  - Bitwise operator
  - Special operators

### 2.11.1 Arithmetic operators

- The arithmetic operators are

| Operator | Meaning |
|---|---|
| + | Addition or unary plus |
| - | Subtraction or unary minus |
| * | Multiplication |
| / | Division |
| % | Modulo division (Remainder) |

ISBN: 978-93-91373-84-9

#### 2.11.1.1 *Integer Arithmetic*

- When both the operands in a single arithmetic expression are integers, the expression is called an integer expression and the operation is called integer arithmetic.
- Integer arithmetic always gives an integer value.
- Example
  - If a=14 and b=4, a+b =18, a-b=10, a*b=56, a/b=3, a%b = 2

#### 2.11.1.2 *Real Arithmetic*

- An operation involving only real operands is called real arithmetic. A real operand may assume values either in decimal or exponential notation.
- Example
  - If a=20.5 and b=6.4, a+b=26.9, a-b=14.1, a*b=131.2, a/b=3.20313, a%b=1.3

#### 2.11.1.3 *Mixed-mode Arithmetic*

- When one operand is real and the other is integer, the expression is called a mixed-mode arithmetic expression.
- If either operand is of the real type, then the other operand is converted to real
- Example
  - If a=15 and b=10.0 then a/b=1.5

### 2.11.2 Relational Operators

- The comparison of values can be done with the help of relational operators.
- Java supports six relational operators

ISBN: 978-93-91373-84-9

| Operator | Meaning |
|---|---|
| < | Is less than |
| <= | Is less than or equal to |
| > | Is greater than |
| >= | Is greater than or equal to |
| == | Is equal to |
| != | Is not equal to |

- A simple relational expression contains only one relational operator is of the following form

  **ae-1 relational operator ae – 2**

- Ae-1 and ae-2 are arithmetic expressions which may be simple constants, variables or combination of them
- When arithmetic expressions are used first the arithmetic expression is evaluated and then the result is compared.
- Arithmetic operators have higher priority over relational operators
- Example
- If a=2,b=3,c=8
  - The expression (a+b)<c would be evaluated as follows
  - First a+b is evaluated ie, 2+3 =5
  - Then this value is compared with c as 5<8 which returns true.

### 2.11.3 Logical Operators

- The logical operators are used when we want to form compound conditions by combining two or more relations.
- Java has three logical operators

ISBN: 978-93-91373-84-9

| Operators | Meaning |
|---|---|
| && | Logical AND |
| \|\| | Logical OR |
| ! | Logical NOT |

- An expression which combines two or more relational expressions is termed as a logical expression or a compound relational expression
- The logical expression gives a true or a false value.
- Example
    - If a=3, b=2,c=1,d=5 then the expression (a<b) && (c<d) returns true.
    - (3<2)&&(1<5)
    - T && T => true
- The && operator returns true when both the operands are true, otherwise false
- The || operator returns true when any one of the operands is true and false when both the operands are false.

### 2.11.4 Assignment Operators

- The assignment operators are used to assign the value of an expression to a variable.
- The assignment operator is "="
- Java has a set of shorthand assignment operators which are used of the form

v op=exp;

- Where v is a variable, exp is an expression and op is a Java binary operator.
- The operator op= is known as the shorthand assignment operator
- Example

ISBN: 978-93-91373-84-9

- a+=1 is equivalent to a=a+1
- a-=2 a= a-2
- a*(n+1) a=a*(n+1)

- The use of shorthand assignment operators has three advantages
  - What appears on the left-hand side need not be repeated and therefore it becomes easier to write
  - The statement is more concise and easier to read
  - Use of shorthand operator results in a more efficient code.

### 2.11.5 Increment and Decrement Operators

- Java has two very useful operators they are called the increment and decrement operators
- The operators are
  - ++ -- increment operator
  - -- -- decrement operator
- The ++ operator adds 1 to the operand while the - operator subtracts 1 to the operand. Both are unary operators and are used in the following form
  - ++ m or m++
  - –m or m–
- A prefix operator first adds 1 to the operand and then the result is assigned to the variable on the left.
- The postfix operator first assigns the value to the variable on the left and then increments the operand.

### 2.11.6 Conditional Operator

- The character pair ?: is a ternary operator available in Java. This operator is used to construct conditional expressions of the form

  exp1 ? exp2 : exp3

ISBN: 978-93-91373-84-9

- Where exp1, exp2 and exp3 are expressions
- The operator ?: works as follows
  - Exp1 is evaluated first, if it is nonzero (True), then the expression exp2 is evaluated and becomes the value of the conditional expression
  - If exp1 is false, exp3 is evaluated and its value becomes the value of the conditional expression
- Example
  - If a=10, b=15 then the expression
  - X=(a > b) ? a : b would return the following value
  - X=15

### 2.11.7 Bitwise Operators

- Java has special operators known as bitwise operators for manipulation of data at values of bit level.
- These operators are used for testing the bits, or shifting them to the right or left.
- Bitwise operators may not be applied to float or double.
- The bitwise operators are

| Operator | Meaning |
|---|---|
| & | Bitwise AND |
| ! | Bitwise OR |
| ^ | Bitwise exclusive OR |
| ~ | Ones complement |
| << | Shift left |
| >> | Shift right |
| >>> | Shift right with zero fill |

ISBN: 978-93-91373-84-9

### 2.11.8 Special Operators

- The special operators are
  - instanceof operator
  - Member selection operator (.)
- Instanceof operator is an object reference operator and returns true if the object on the left-hand side is an instance of the class given on the right hand side
- Example:
  - Person instanceof Student - is true if the object person belongs to the class student.
- Dot operator is used to access the instance variables and methods of class objects.
- Example
  - Person.age
  - Person.getdata()

## 2.12 Evaluation of Expressions

- Expressions are evaluated using an assignment statement of the form
  - Variable = expression
- Variable is any valid Java variable name. When the statement is encountered the expression is evaluated first and the result then replaces the previous value of the variable on the left-hand side.
- Example
  - X=a*b-c
  - Y=b/c+a

ISBN: 978-93-91373-84-9

## 2.13 Type Conversions in Expressions

### 2.13.1 Automatic Type Conversion

- Java permits mixing of constants and variables of different types in an expression, but during evaluation strict rules are followed
- If the operands are of different types, the "lower" type is automatically converted to the "higher" type before the operation proceeds. The result is of higher type.
- The following changes are introduced during the final assignment
  - Float to int causes truncation of the fractional part
  - Double to float causes rounding of digits
  - Long to int causes dropping of the excess higher order bits.

### 2.13.2 Casting a Value

- There are instances when we want to force a type conversion
- The process of local conversion is called as casting a value.
- Syntax
  - (type_name)expression
- Where type_name is one of the standard data types. The expression may be a constant, variable or an expression
- Example:
  - X=(int)7.5
  - A=(int) 21.3/(int) 4.5

### 2.13.3 Generic Type Casting

- Generics eliminate the need of explicit type casting in collection. A collection is a set of interfaces and classes that sort and manipulate a group of data into a single unit.

ISBN: 978-93-91373-84-9

- To retrieve elements from a collection, we need to typecast the elements, because each element is considered as an object in collection

## 2.14 Operator Precedence and Associativity

- Precedence is used to determine how the expression involving more than one operator is evaluated.
- The operator at the higher level of precedence is evaluated first.
- The operator with the same precedence is either evaluated from left to right or right to left. This is known as the associativity of the operator.

| Operator | Description | Associativity | Rank |
|---|---|---|---|
| . | Member selection | Left to right | 1 |
| ( ) | Function call | | |
| [ ] | Array element reference | | |
| - | Unary minus | Right to left | 2 |
| ++ | Increment | | |
| -- | Decrement | | |
| ! | Logical negation | | |
| ~ | Ones complement | | |
| (type) | Casting | | |
| * | Multiplication | Left to Right | 3 |
| / | Division | | |
| % | Modulus | | |
| + | Addition | Left to Right | 4 |
| - | Subtraction | | |
| << | Left Shift | Left to Right | 5 |
| >> | Right Shift | | |

ISBN: 978-93-91373-84-9

<table>
<tr><td>>>></td><td>Right Shift with zero fill</td><td></td><td></td></tr>
<tr><td><</td><td>Less than</td><td rowspan="5">Left to Right</td><td rowspan="5">6</td></tr>
<tr><td><=</td><td>Less than or equal to</td></tr>
<tr><td>></td><td>Greater than</td></tr>
<tr><td>>=</td><td>Greater than or equal to</td></tr>
<tr><td>Instanceof</td><td>Type comparison</td></tr>
<tr><td>==</td><td>Equality</td><td rowspan="2">Left to Right</td><td rowspan="2">7</td></tr>
<tr><td>!=</td><td>Inequality</td></tr>
<tr><td>&</td><td>Bitwise AND</td><td rowspan="5">Left to Right</td><td>8</td></tr>
<tr><td>^</td><td>Bitwise XOR</td><td>9</td></tr>
<tr><td>|</td><td>Bitwise OR</td><td>10</td></tr>
<tr><td>&&</td><td>Logical AND</td><td>11</td></tr>
<tr><td>| |</td><td>Logical OR</td><td>12</td></tr>
<tr><td>?:</td><td>Conditional operator</td><td rowspan="2">Right to Left</td><td>13</td></tr>
<tr><td>=</td><td>Assignment operator</td><td>14</td></tr>
<tr><td>Op=</td><td>Shorthand assignment</td><td></td><td></td></tr>
</table>

ISBN: 978-93-91373-84-9

# Chapter 3
# Decision Making and Looping Statements

## 3 Introduction

- When a program breaks the sequential flow and jumps to another part of the code, it is called branching
- When branching is based on a particular condition, it is known as conditional branching.
- If branching takes place without any decision, it is known as unconditional branching
- Java supports the following statements known as control or decision making statements
  - If statement
  - Switch statement
  - Conditional operator statement

### 3.1.1 Decision Making with if Statement

- The if statement is a powerful decision making statement and is used to control the flow of execution of statements
- It is basically two-way decision statement.
- The syntax
  - If(test expression)
- The test expression is evaluated first. And based on the value of the expression if true or false control is transferred to a particular statement
- It has two paths to follow, one for the true condition and the other for the false condition.

ISBN: 978-93-91373-84-9

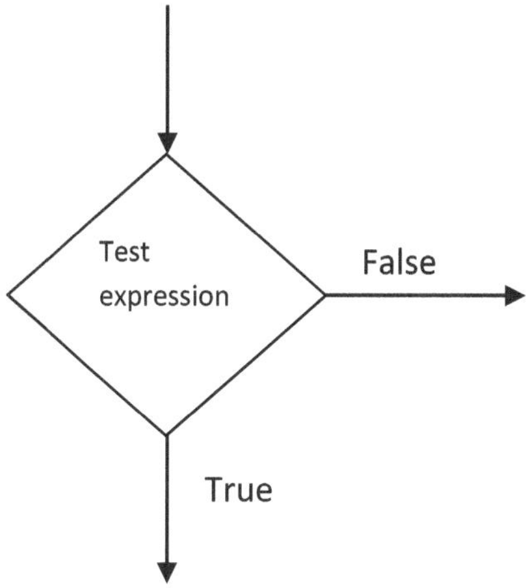


### 3.1.1.1 *Simple if Statement*

- **Syntax**

```
if(test expression)
{
        Statement-block;
}
Statement-x;
```

- The test expression is evaluated
  - If true, the statement-block is executed and if false the statement x is evaluated.
  - If the expression is true, both the statement-block and the statement-x is executed

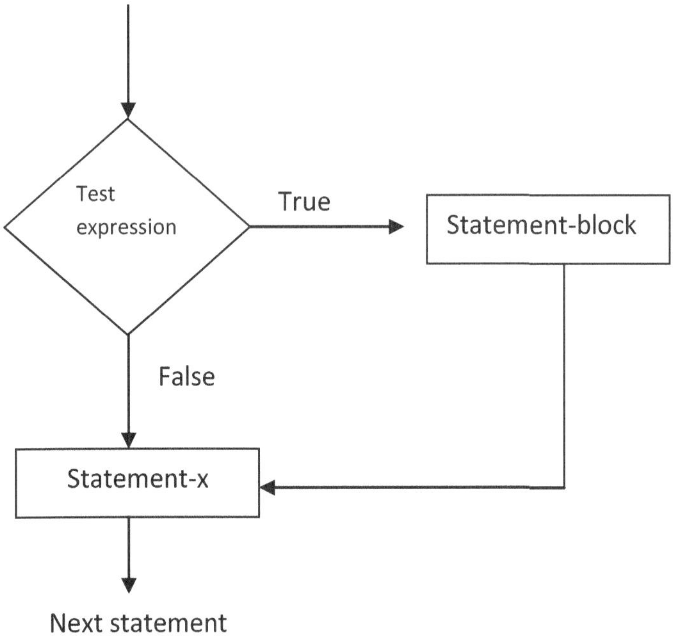


- **Example: Program to check whether the person meets the sports criteria**

```
import java.io.*;
class ex1
{
    public static void main(String a[])throws IOException
    {
      String name,cat;
      DataInputStream in=new DataInputStream(System.in);
      System.out.print("Enter the name: ");
      name=in.readLine();
      System.out.print("Enter the category:");
      cat=in.readLine();
      if (cat=="SPORTS")
      {
```

ISBN: 978-93-91373-84-9

```
            System.out.println(name+" is a sports person");
          }
          System.out.println("Name: "+name);
          System.out.println("Category: "+cat);
        }
}
```

**Output**

```
Enter the name: raj
Enter the category: SPORTS
Raj is a sports person
Name: raj
Category: SPORTS
```

### 3.1.2 The If...Else Statement

- **The general form is**

```
if(test expression)
{
        True-block statement;
}
else
{
        False-block statement;
}
Statement-x;
```

- The test expression is evaluated
  - If the expression is true, the true-block statement is executed
  - If the expression is false, the false-block statement is executed.
  - In either of the case the statement-x is executed.

ISBN: 978-93-91373-84-9

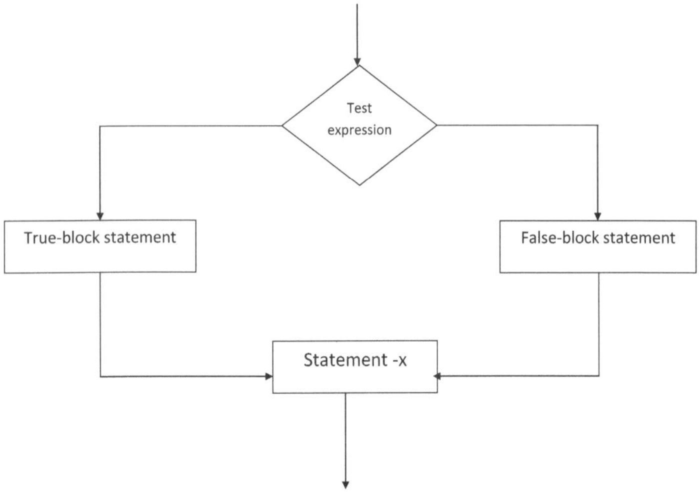


- **Example: Program to check whether a given number is odd or even**

```
import java.io.*;
class eve
{
    public static void main(String a[])throws IOException
    {
      DataInputStream in = new DataInputStream(System.in);
      int a,even=0,odd=0;
      System.out.println("Enter the number: ");
      a=Integer.parseInt(in.readLine());
      if(a%2==0)
        System.out.println(a+" is even");
      else
        System.out.println(a+" is odd");
    }
}
```

**Output:**

```
Enter the number: 17
17 is odd.
```

ISBN: 978-93-91373-84-9

### 3.1.3 Nesting of If...Else statement

- When a series of decisions are involved, we have to use more than one if...else statement in nested form
- **Syntax**

```
if (condition1)
{
        if(condition2)
        {
                Statement1;
        }
        else
        {
                Statement2;
        }
}
else
{
        Statement3;
}
Statementx;
```

- Execution
    - If condition1 is false, the statment3 will be executed.
    - If condition1 is true, it continues to perform the second test. If condition2 is true, the statement1 is executed otherwise statement2 is executed. And finally statement is executed.

ISBN: 978-93-91373-84-9

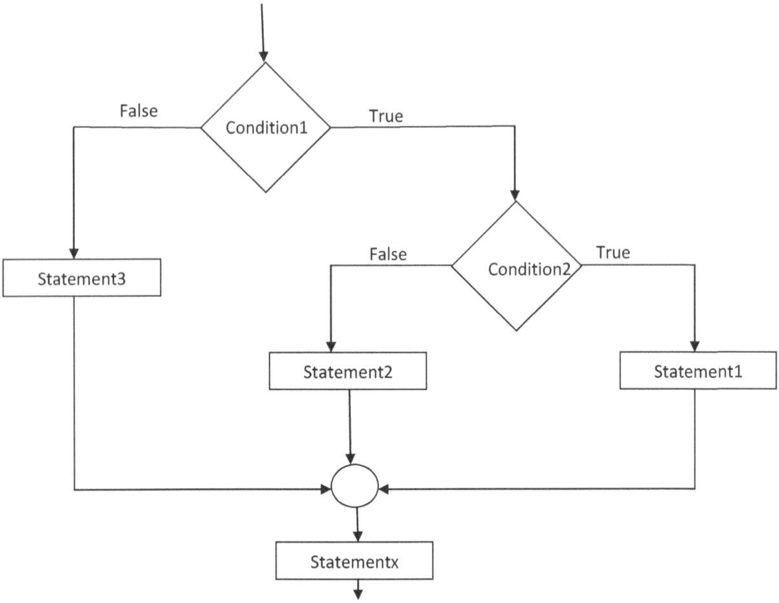


- **Example: Program to check the greatest of three numbers**

```
import java.io.*;
class great
{
    public static void main(String a[])throws IOException
    {
      DataInputStream in=new DataInputStream(System.in);
      int a,b,c;
      System.out.print("Enter the value for a: ");
      a=Integer.parseInt(in.readLine());
      System.out.print("Enter the value for b: ");
      b=Integer.parseInt(in.readLine());
      System.out.print("Enter the value for c: ");
      c=Integer.parseInt(in.readLine());
      if(a>b)
      {
        if(a>c)
```

ISBN: 978-93-91373-84-9

```
        {
                System.out.println(a+" is greatest");
        }
        else
        {
                System.out.println(c+" is greatest");
        }
      }
      else
      {
        if(b>c)
        {
                System.out.println(b+" is greatest");
        }
        else
        {
                System.out.println(c+" is greatest");
        }
      }
    }
}
```

**Output:**

```
Enter the value for a: 12
Enter the value for b: 6
Enter the value for c: 190
190 is greatest
```

ISBN: 978-93-91373-84-9

### 3.1.4 The Else..if Ladder

- The other way of writing ifs together when multipath decisions are involved is called the else if ladder

**Syntax**

```
if(condition1)
        statement1;
else if(condition2)
        statement2;
else if(condition3)
        statement3
...............
else if(condition)
        statementn;
else
        default-statement;
statement-x;
```

- **Execution**
  - If the condition1 is true, the statement1 is executed and the control is transferred to statement
  - If the condition1 is false, then the condition2 is tested and the evaluation happens.
  - If all the conditions are false, then the default statement is executed and then the control is transferred to statementx

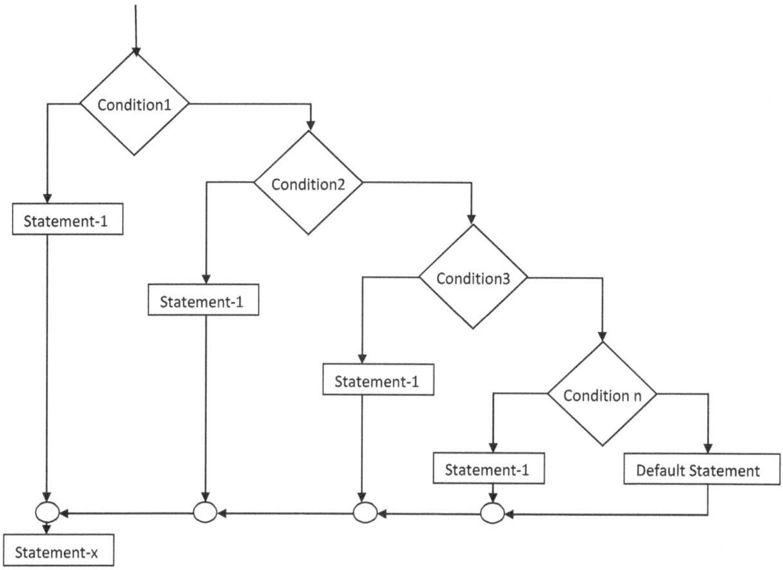


- **Example: Program to find the grade of students depending on marks**

```
import java.io.*;
class grade
{
    public static void main(String a[])throws IOException
    {
      DataInputStream in=new DataInputStream(System.in);
      int mks;
      System.out.println("Enter the marks:");
      mks=Integer.parseInt(in.readLine());
      if(mks>79)
        System.out.println("Grade = Honours");
      else if(mks>59)
        System.out.println("Grade = First Division");
      else if(mks>49)
        System.out.println("Grade = Second Division");
      else if(mks>39)
```

ISBN: 978-93-91373-84-9 

```
            System.out.println("Grade = Third Division");
          else
            System.out.println("Grade = Fail");
      }
}
```

**Output:**

Enter the marks: 55
Grade = Second Division

### 3.1.5 Switch Statement

- The complexity of using if..else increases when the number of alternatives increases. The program becomes difficult to read and follow.
- Java has a built-in multiway decision statement known as switch. The switch statement tests the value of a given variable against a list of case values and when a match is found, a block of statement associated with that case is executed.
- **Syntax**

```
switch(expression)
{
        case value-1:   block-1;
                        break;
        case value-2:   block-2;
                        break;
        ....................
        ..................
        default:        defalult-block;
}
Statement-x;
```

ISBN: 978-93-91373-84-9

- The expression is an integer expression or characters.
- Value-1, value-2 are constants or constant expressions and are called as case labels. Each of these values should be unique within a switch statement.
- Block-1, block-2 are statement lists and may contain zero or more statements. There is no need to put braces around these blocks but the case labels should end with a colon(:)
- The break statement at the end of each block signal causes an exit from the switch statement, transferring the control to statement-x following the switch.
- The default is an optional case.
- The flow of control is as follows

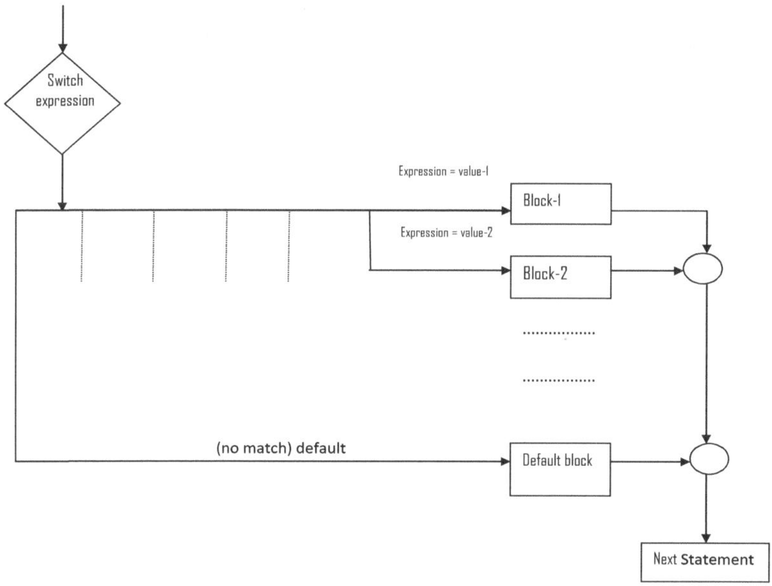


- **Example: Program to print the grade of the students**

```
import java.io.*;
class ggrd
{
    public static void main(String a[])throws IOException
```

ISBN: 978-93-91373-84-9

```
{
  int mks,n;
  DataInputStream in=new DataInputStream(System.in);
  System.out.println("Enter the marks: ");
  mks=Integer.parseInt(in.readLine());
  n=mks/10;
  switch(mks)
  {
    case 10:
    case 9:
    case 8:        System.out.println("Grade=Honours");
                   break;
    case 7:
    case 6:        System.out.println("Grade=First
Division");
                   break;
    case 5:        System.out.println("Grade=Second
Division");
                   break;
    case 4:        System.out.println("Grade=Third
Division");
                   break;
    case 3:        System.out.println("Grade=Fail");
                   break;
    default:       System.out.println("Grade=Fail");
                   break;
  }
 }
}
```

**Output:**

Enter marks: 67

Grade=First Division

## 3.2 Decision Making and Looping

- The process of repeatedly executing a block of statements is known as looping.
- The statements in the block may be executed any number of times, from zero to infinite. If the loop continues forever, it is called as infinite loop
- In looping, a sequence of statements is executed until some conditions for the termination of the loop are satisfied.
- The program loop therefore consists of two segments
  - Body of the loop
  - Control statement
- The control statement tests the conditions and then directs the repeated execution of the statements contained in the body of the loop
- Depending on the position of the control statement in the loop, a control structure may be classified into two
  - Entry-controlled loop
  - Exit-controlled loop
- In an entry-controlled loop the condition is first tested before the start of the loop execution. If the condition is false the loop body is not executed even once.
- In an exit-controlled loop the condition is placed at the end of the loop body so the body of the loop is executed atleast once before the condition is tested.
- A looping process includes the four steps
  - Setting and initialization of a counter
  - Execution of the statements in the loop

ISBN: 978-93-91373-84-9

  - Test for a specified condition for execution of the loop
  - Incrementing the counter
- The test is to determine whether the loop has repeated for a specified number of times.
- Java provides three constructs
  - While
  - Do...while
  - For

### 3.2.1 The While Statement

- The simplest of the looping structure is the while statement.
- **Syntax**

```
Initialization;
while(test condition)
{
        Body of the loop;
}
```

- The while loop is an entry-controlled loop statement.
- **Execution**
  - The test condition is evaluated, and if the condition is true, the body of the loop is executed.
  - After the body of the loop is executed, the test condition is again evaluated and if true the body is again executed otherwise the loop is terminated.
  - The body of the loop may have one or more statements and the braces are necessary if the body contains two or more statements.
- **Example: Program to print the sum of the squares of the given numbers $1^2+2^2+3^2+....+n^2$**

```
import java.io.*;
class sum
```

ISBN: 978-93-91373-84-9

```
{
    public static void main(String a[]) throws IOException
    {
      DataInputStream in=new DataInputStream(System.in);
      int n,i,sum=0;
      System.out.println("Enter the number of terms: ");
      n=Integer.parseInt(in.readLine());
      i=1;
      while(i<=n)
      {
        sum=sum+i*i;
        i++;
      }
      System.out.println("Sum of series = "+sum);
    }}
```

**Output:**

```
Enter the number of terms:  8
Sum of series =
```

### 3.2.2 The do Statement

- In some situations the body of the loop must be executed before the condition is tested. This is performed using the do statement
- **Syntax**

```
Initialization
do
{
        Body of the loop
}while(condition);
```

ISBN: 978-93-91373-84-9 

- **Execution**
  - The body of the loop is first executed and then the condition is tested in the while statement.
  - If the condition is true, the body of the loop is executed otherwise the loop is terminated.
- **Example: Program to display the Fibonacci series using do... while loop**

```
import java.io.*;
class fibo
{
    public static void main(String a[])throws IOException
    {
      int n,f1=-1,f2=1,f,i;
      DataInputStream in=new DataInputStream(System.in);
      System.out.println("Enter the number of terms: ");
      n=Integer.parseInt(in.readLine());
      System.out.println("Fibonacci Series");
      i=1;
      do
      {
        f=f1+f2;
        System.out.print(f+"\t");
        f1=f2;
        f2=f;
        i++;
      }while(i<=n);
    }
}
```

**Output:**

```
Enter the number of terms:5
0  1       2       3       5
```

ISBN: 978-93-91373-84-9

### 3.2.3 The for Statement:

- The for loop is a entry controlled loop that provides a more concise loop control structure.
- **Syntax**

```
for(initialization;test condition;increment)
{
        Body of the loop
}
```

- **Execution**
  - Initialization of the control variable is done using the assignment statement first. The control variable is also called as loop-control variable.
  - The value of the control variable is tested with the test condition. If the condition is true the body of the loop is executed otherwise the loop is terminated and control is transferred outside the loop
  - After the body of the loop is executed, control is transferred back to the for statement with the control variable incremented. Again the test condition is checked and then the body of the loop is executed .
- **Example: Program to print the sum of the squares of the given numbers $1^2+2^2+3^2+....+n^2$Using for loop**

```
import java.io.*;
class sum1
{
    public static void main(String a[])throws IOException
    {
      int i,n,sum=0;
      DataInputStream in=new DataInputStream(System.in);
      System.out.println("Enter the number of terms: ");
```

ISBN: 978-93-91373-84-9

```
        n=Integer.parseInt(in.readLine());
        for(i=1;i<=n;i++)
        {
          sum=sum+i*i;
        }
        System.out.println("Sum of terms = "+sum);
    }
}
```

**Output:**

```
Enter the number of terms: 8
Sum of terms = 36
```

### 3.2.3.1 Additional Features of for Loop

- In the initialization clause more than one variable can be initialized separated by comma. The increment clause can also have more than one variable separated by comma
- Example

```
for(i=1,j=1;i<=n;i++,j++)
{                    }
```

- Another unique feature of for loop is that one or more sections can be omitted.
- Example

```
m=5;
for( ; m!=100; )
{
        System.out.println(m);
        m+=5;
}
```

- In the above example, the initialization is done outside the loop and the increment is done within the body of

ISBN: 978-93-91373-84-9

the loop. In such situations the clause must be left empty and separated by a semicolon.

- We can create time delay loops using null statements
- Example
  for(i=1;i<=100;i++)
  ;

### 3.2.4 Nesting of for Loops

- Nesting of for loop is allowed in java, ie one for loop can be written within another for loop.
- Example

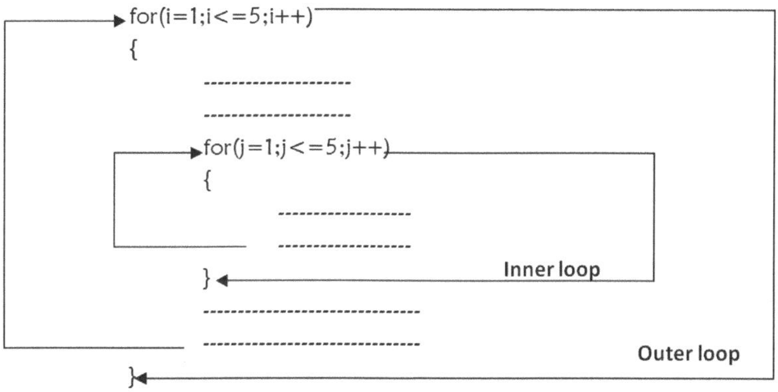


- Some points should be noted when using nesting of loops
  - The inner loop should be completely within the outer loop.
  - The control variables of both the loops should be different.
  - The inner loop should be completed first and then the outer loop.

ISBN: 978-93-91373-84-9

## 3.3 Jumps in Loops

- Sometimes, when executing a loop it becomes necessary to skip a part of the loop or to leave the loop as soon as a certain condition occurs.
- In such cases we use jumps in loops
- Java permits a jump from one statement to the end or beginning of a loop as well as a jump out of a loop

### 3.3.1 Jumping out of a Loop

- We use two statements within the loop
  - Break
  - Continue
- Break statement is used to exit from a loop and the continue statement is used to skip the statements in a loop and continue with the next iteration

- **Example: Program to generate prime numbers**

```
class pr
{
    public static void main(String a[])
    {
      int n=15,j,i,c=0;
      System.out.println("1");
      System.out.println("2");
      System.out.println("3");
      for(i=1;i<=50;i++)
      {
        for(j=2;j<=i/2;j++)
        {
                if(i%j==0)
                {
                        c=0;
```

ISBN: 978-93-91373-84-9

```
                        break;
                }
                else
                {
                        c=1;

        continue;
                }

          }
          if(c==1)
                System.out.println(i);
        }
    }
}
```

**Output:**

1 2 3 5 7 11 13

# CHAPTER 4
# Classes, Objects and Methods

## 4.1. Classes, Objects and Methods

- Java is a true object-oriented language and so everything in java should be written within classes.
- Java programs must be encapsulated in a class that defines the state and behaviour of the basic program components known as objects.
- Classes create objects and objects use methods to communicate between them.
- Classes provide a convenient method for packing together a group of logically related data items and functions that work on them.
- In java data items are called fields and functions are called methods.
- Calling a method in an object is described as sending the object a message.

## 4.2. Defining a class

- Classes are the basic building blocks of Java programs. Classes are called as Abstract Data Type (ADT).
- Classes describe the structure of the object. Java classes consist of both attributes and behaviours.
- Attributes represent the data that is unique to an instance of a class, while behaviours are methods that operate on the data to perform useful tasks
- **Syntax**
  [public][(abstract | final)]class classname [extends superclassname]
  {

```
        [field declarations;]
        [methods declaration;]
}
```

- A class can have public or default (no modifier) visibility.
- It can be either abstract, final or with no modifier.
- It must have the class keyword, and class must be followed by a legal identifier.
- It may optionally extend one parent class. By default, it will extend java.lang.Object.
- It may optionally implement any number of interfaces separated by commas.

### *4.2.1. Fields Declaration*

- Data is encapsulated in a class by placing the data fields inside the body of the class definition.
- These variables are called instance variables because they are created whenever an object is instantiated.
- Example

```
class Rectangle
{
        int len, breadth;
}
```

### *4.2.2. Methods Declaration*

- A class with only data fields has no life. Because these objects cannot respond to messages.
- So we need to add methods for manipulating the data in the class. Methods are declared inside the body of the class.
- **Syntax**

```
type methodname(parameter-list)
{
```

```
        Method body;
}
```

- Method declaration has four parts
  - The name of the method (methodname) - could be any valid identifier
  - The type of value the method returns (type) - the value could be any datatype like int, float etc
  - A list of parameters (parameter-list) - should always be enclosed within parenthesis
  - The body of the method - the operations to be performed on the data
- Example

```
class Rect
{
        int len,bre,area;
        void get()
        {
                len=23;
                bre=10;
        }
        void calc()
        {
                area=len*bre;
        }
}
```

## 4.3. Creating Objects

- An object in java is a block of memory that contains space to store all the instance variables.
- Creating an object is called as instantiating an object.
- Objects of different classes can be created, used and destroyed in the course of executing a program.

- Objects in java are created using the "new" operator
- **Syntax**

```
classname objectname;
        //declare the object
Objectname=new classname();
        //instantiate the object
```

- The above two line can be combined and written in one statement as follows
    - classname objectname = new classname();
- Example
    - rect obj= new rect();

### 4.3.1. Accessing Class Members

- Each object has its own set of variables, and these variables should be assigned values to use them in the program.
- Every variables must be assigned values before they are used
- The objects are used to access the variables from outside the class. To access the class member variables the dot operator is used.
- **Syntax**

```
objectname.variablename=value;
objectname.methodname(parameter-list);
```

- The objectname is the name of the object, and the variablename is the name of the instance variable, methodname is the name of the method we wish to call, parameter list is a list of actual values separated by comma
- **Example: Program to show the application of class,objects and methods**

```
class Calculator
```

ISBN: 978-93-91373-84-9

```
{
        int n1,n2,res;
        void setData(int x1,int x2)
        {
                n1=x1;
                n2=x2;
        }
        void add()
        {
                res=n1+n2;
                System.out.println("The Sum is : "+res);
        }
        void sub()
        {
                res=n1-n2;
                System.out.println("The Difference is : "+res);
        }
        int mul()
        {
                res=n1*n2;
                return res;
        }
        double div()
        {
                return n1/n2;
        }
}
class Mainprg
{
        public static void main(String args[])
        {
                Calculator cal=new Calculator();
```

ISBN: 978-93-91373-84-9

```
        cal.setData(20,10);
        cal.add();
        cal.sub();
        int y=cal.mul();
        System.out.println("The Product is : "+y);
        double d=cal.div();
        System.out.println("The divided value is :"+d);

    }
}
```

## 4.4. Constructors

- Java supports a special type of method, called a constructor that enables an object to initialize itself when it is created.
- Constructors will have the same name as that of the class. Constructors and methods differ in two important aspects of the signatures. They are
    - Modifiers
    - Return types
- Like methods, constructors can have any of the access modifiers: public, protected, private, or none (often called friendly). However constructors cannot be made abstract, final, native, static, or synchronized.
- Methods can have any valid return type, or no return type, in which case the return type is given as void. Constructors cannot take any return type, even void.
- **Example: The Calculator class created in example 1 is modified, the setData() method is replaced with a constructer.**

```
class Calculator
{
```

ISBN: 978-93-91373-84-9

```
        int n1,n2,res;
        Calculator (int x1,int x2) //Constructor
        {
                n1=x1;
                n2=x2;
        }
        void add()
        {
                res=n1+n2;
                System.out.println("The Sum is : "+res);
        }
        void sub()
        {
                res=n1-n2;
                System.out.println("The Difference is : "+res);
        }
        int mul()
        {
                res=n1*n2;
                return res;
        }
        double div()
        {
                return n1/n2;
        }
}
class Mainprg
{
        public static void main(String args[])
        {
                Calculator cal=new Calculator(20,10); //Calls
        the constructer
```

```
            cal.add();
            cal.sub();
            int y=cal.mul();
            System.out.println("The Product is : "+y);
            double d=cal.div();
            System.out.println("The divided value is :"+d);
    }
}
```

**Output:**

```
The Sum is : 30
The Difference is : 10
The Product is : 200
The divided value is : 2.0
```

## 4.5. Method Overloading

- In Java it is possible to create methods that have the same name, but different parameter lists and different definitions. This is called method overloading
- In Java two or more methods can share the same name as long as either the type of their arguments differs or the number of their arguments differs - or both. When two or more methods share the same name, they are said *overloaded*.
- Method Overloading is used when objects are required to perform similar tasks but using different input parameters.
- When we call a method using an object, Java matches the method name first and then the number and type of parameters to decide which one of the definitions to execute. This process is known as polymorphism

ISBN: 978-93-91373-84-9

- Example: The coding listed below overloads the method sum()

```
class numbers
{
    void sum(int x,int y)
    {
            int z=x+y;
        System.out.println("Sum of addition of two
    numbers : "+z);
    }
    void sum(int x,int y,int z)
    {
            int r=x+y+z;
        System.out.println("Sum of addition of three
    numbers : "+r);
    }
}
class test
{
    public static void main(String args[])
    {
        numbers t=new numbers();
        t.sum(10,20); //Call the method sum with two
    parameters
        t.sum(10,20,30); //Call the method sum with three
    parameters
    }
}
```

**Output:**

Sum of addition of two numbers : 30

Sum of addition of three numbers : 60

ISBN: 978-93-91373-84-9

## 4.6. Constructor Overloading

- In Java, similar to the concept of method overloading we have constructor overloading. In constructor overloading we have multiple constructors that differ either in type of their arguments or the number of their arguments or both.
- The constructor that takes no parameter is called default constructor. Below is the example of constructor overloading.
- Example: The program given below makes use of constructor overloading concept.

```
class dimension
{
        int x1,y1;
        dimension() //default constructor
        {
                x1=0;
                y1=0;
        }
        dimension(int a)
        {
                x1=a;
                y1=0;
        }
        dimension(int a,int b)
        {
                x1=a;
                y1=b;
        }
        void display()
        {
                System.out.println(" x1 is :"+x1);
```

ISBN: 978-93-91373-84-9

```
		System.out.println(" y1 is :"+y1);
	}
}
class Mainprg
{
	public static void main(String args[])
	{
		dimension d1=new dimension(10,50); //Calls
	constructer with twoparameter
		dimension d2=new dimension(10); //Calls
	constructer with oneparameter
		dimension d3=new dimension(); //Calls
	default constructer
		d1.display();
		d2.display();
		d3.display();
	}
}
```

**Output**

```
X1 is : 10    Y1 is : 50    X1 is : 10    Y1 is : 0
X1 is : 0     Y1 is : 0
```

## 4.7. Static Members

- Sometimes it is desirable for all the objects of a particular class to share data. In Java this can be achieved by defining the corresponding class members as static using the keyword static.
- Since static variables are common to all the instances of the class, static variables can also be called as class variables.

ISBN: 978-93-91373-84-9

- For a static data member there will be only one copy of the class member shared by all objects of the class.
- A static data member is like a global variable, but has class-level scope.
- To declare a variable as static we have to simply prefix the variable declaration with the keyword static.
- For example
  - static int x;

### 4.7.1. Static Methods

- Like static variables, static methods are called without the object. The differences between a static methods and non-static methods are as follows.
  - A static method can access only static data and can call only other static methods. A non-static member function can access all of the above including the static data member.
  - A static method can be called, even when a class is not instantiated, a non-static member function can be called only after instantiating the class as an object.
  - A static method cannot have access to the 'this' keyword.
  - A static method cannot have access to the 'super' keyword.
- **Example: This program makes use of static data members and methods**

```
class sta
{
	static int increment(int k)
	{
		k++;
```

ISBN: 978-93-91373-84-9

```
            return k;
        }
        static int decrement(int j)
        {
            j–;
            return j;
        }
}
class static_test
{
        public static void main(String args[])
        {
            int a=static_test.increment(10);
            System.out.println("Incremented Value is :
        "+a);
            int b=static_test.decrement(20);
            System.out.println("Decremented Value is :
        "+b);
        }
}
```

In Java class libraries many static methods are found. For example in java.langpackage, we have a class called "Math" all the methods declared in the Mathclass are static.

## 4.8. Nesting of Methods

- The method of a class can be called only by an object of that class using the dot operator. In java a method can be called by using only its name by another method of the same class this is called nesting of methods.
- Example: Program to show nesting of methods

```
class nes
{
```

```
    int m,n;
    nes(int x, int y)
    {
      m=x;
      n=y;
    }
    int largest()
    {
      If(m>n)
        return(m);
      else
        return(n);
    }
    void display()
    {
      int large=largest();
      System.out.println("Largest value = "+large);
    }
}
class nest
{
    public static void main(String a[])
    {
      nes  obj = new nes(23,4);
      obj.display();
    }
}
```

ISBN: 978-93-91373-84-9

# CHAPTER 5
# Inheritance

## 5. Inheritance: Extending a Class

- Inheritance is a feature by which new classes are derived from the existing classes.
- Inheritance allows re-using code without having to rewrite from scratch.
- Inheritance is a compile-time mechanism in Java that allows you to extend a class (called the base class or superclass) with another class (called the derived class or subclass).
- In Java, inheritance is used for two purposes:
  - Class inheritance - create a new class as an extension of another class, primarily for the purpose of code reuse. That is, the derived class inherits the public methods and public data of the base class. Java only allows a class to have one immediate base class, i.e., single class inheritance
  - Interface inheritance – Interface inheritance is used to achieve multiple inheritance.

## 5.1. Defining a subclass

```
class subclassname extends superclassname
{
        Variable declaration;
        Methods declaration;
}
```

- The keyword extends signifies that the properties of the superclassname are extended to the subclassname.

ISBN: 978-93-91373-84-9 

- In Java the following forms of class inheritance is supported.
  - Single Inheritance – only one super class
  - Multiple Inheritance – several super classes
  - Multilevel Inheritance – derived from a derived class.
  - Hierarchical Inheritance - one super class, many subclasses
- Example: Program to show single inheritance

```
class rect
{
    int l,b;
    rect(int x, int y)
    {
      l=x;
      b=y;
    }
    void rect_area()
    {
      Sytsem.out.println("Area of Rectangle = "+(l*b));
    }
}
class cube extends rect
{
    int h;
    cube(int x, int y, int z)
    {
      super(x,y);
      h=z;
    }
        void cube_vol()
    {
```

ISBN: 978-93-91373-84-9

```
        Sytem.out.println("Volume of the cube = "+(l*b*h));
    }
}
class mainprg
{
    public static void main(String a[])
    {
        cube c= new cube(10,20,30);
        c.rect_area();
        c.cube_vol();
    }
}
```

**Output:**

```
Area of rectangle = 200
Volume of cube = 6000
```

### 5.1.1. Subclass Constructor / Super Keyword

- A subclass constructor is used to construct the instance of variables of both the subclass and the super class.
- Java uses a unique keyword called "super" to call the super class constructor.
- The syntax is
    - super() or
    - super(parameter list)
- The keyword super is used subject to the following conditions
    - Super may be used on within a subclass constructor method
    - The call to the superclass constructor must appear as the first statement within the subclass constructor

ISBN: 978-93-91373-84-9

    - The parameters in the super call match the order and type of the instance variables declared in the superclass.
- Thus, we can use super keyword in the following three contexts
    - To call a superclass constructer
    - To call a superclass method overridden by the subclass method
    - To refer the superclass variable from the subclass.

## 5.2. Types of Inheritance

- Inheritance is a mechanism of acquiring the features and behaviours of a a class by another class. The class whose members are inherited is called the base class and the class that inherits the members is called the derived class.
- Object Oriented Programming supports six different types of inheritance. They are
    - Single inheritance
    - Multi-level inheritance
    - Multiple inheritance
    - Multipath inheritance
    - Hierarchical Inheritance
    - Hybrid Inheritance

### 5.2.1. Single inheritance

- In single inheritance the derived class is created from a single base class. Single inheritance is depicted in the diagram below.

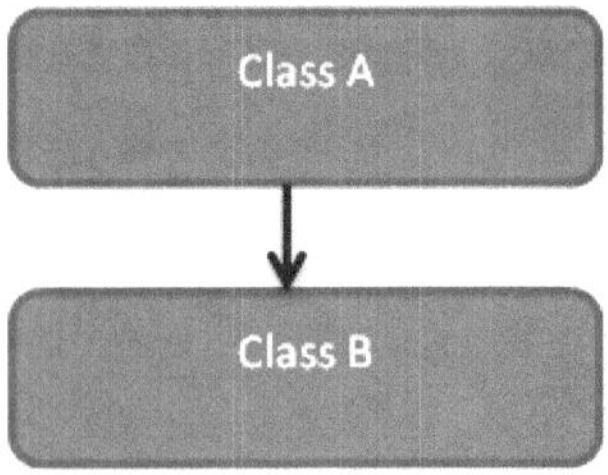


- Here Class A is the parent class and Class B is the derived class and Class B inherits all the features and behaviour of the parent Class A
- Syntax for Single Inheritance

```
//Base Class
class A
{
  public void foo1()
  {
  // to do
  }
}
//Derived Class
class B : A
{
  public void foo2()
  {
        //to do
  }
}
```

### 5.2.2. Multi-level Inheritance

- In Multi-level Inheritance, a derived class is created from another derived class. The multi-level inheritance is depicted in the diagram below

ISBN: 978-93-91373-84-9

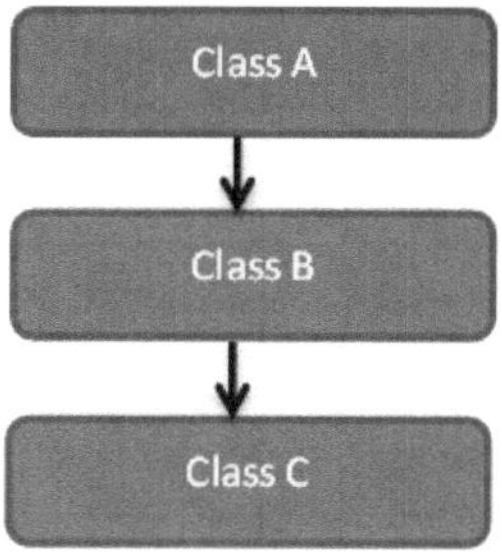


- In the above diagram Class A is the parent class of B and Class B is the parent class of Class C. Class B inherits the properties and behaviour of Class A and Class C implicitly inherits the properties and behaviour of Class A and Class B. Class C is created by inheriting the properties of Class B which in turn is derived from Class A
- Syntax for Multi-level Inheritance

```
//Base Class
class A
{
public void foo1()
{
  //to do
}
}
//Derived Class
class B : A
{
  public void foo2()
  {
          //to do
  }
}
//Derived Class
```

ISBN: 978-93-91373-84-9

```
class C : B
{
  public void foo3()
  {
          //to do
  }
}
```

### 5.2.3. Multiple Inheritance

- In multiple inheritance, a derived class is created from more than one base class. This type of inheritance is shown in the diagram below

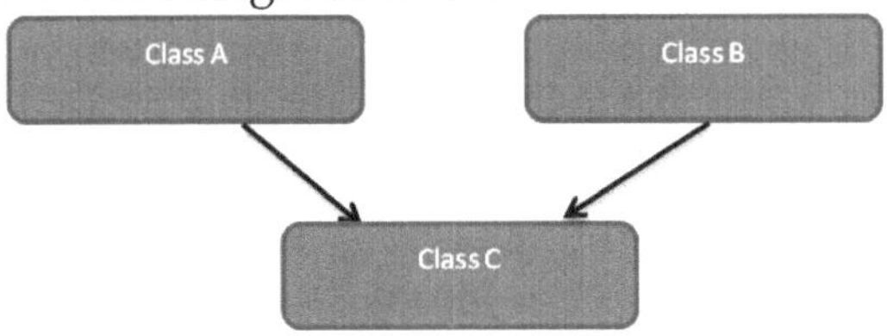


- In the above diagram, Class C is the derived class which inherits the properties and behaviour from two Base Classes class A and Class B.
- Syntax for Multiple Inheritance

```
//Base Class
class A
{
  public void foo1()
  {
          //to do
  }
}
class B
{
  public void foo2()
  {
```

ISBN: 978-93-91373-84-9

```
            //to do
    }
}
//Derived Class
class C : B, A
{
    public void foo3()
    {
            //to do
    }
}
```

### 5.2.4. Multipath inheritance

- In this inheritance, a derived class is created from another derived classes and the same base class of another derived classes.

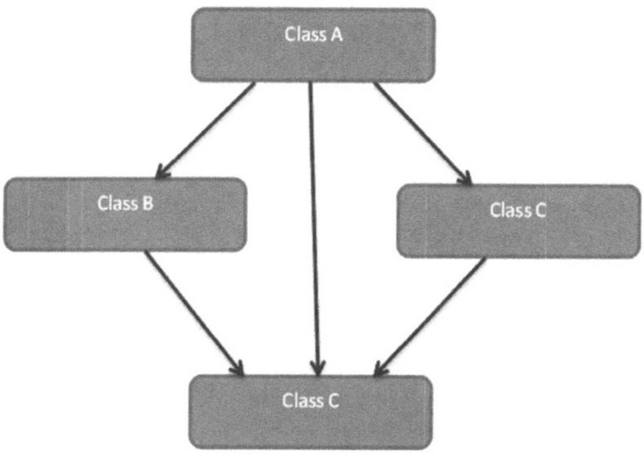


- In the given example, class D inherits the properties and behavior of class C and class B as well as Class A. Both class C and class B inherits the Class A. So, Class A is the parent for Class B and Class C as well as Class D. So it's making it Multipath inheritance.
- Syntax for Multipath Inheritance

```
//Base Class
class A
```

ISBN: 978-93-91373-84-9

```
{
  public void foo1()
  {
          //to do
  }
}
//Derived Class
class B : A
{
  public void foo2()
  {
          //to do
  }
}
//Derived Class
class C : A
{
  public void foo3()
  {
          //to do
  }
}
class D : C,A,B
{
  public void foo4()
  {
          //to do
  }
}
```

### 5.2.5. Hierarchical Inheritance

- In this inheritance, more than one derived classes are created from a single base class and further child classes act as parent classes for more than one child classes.

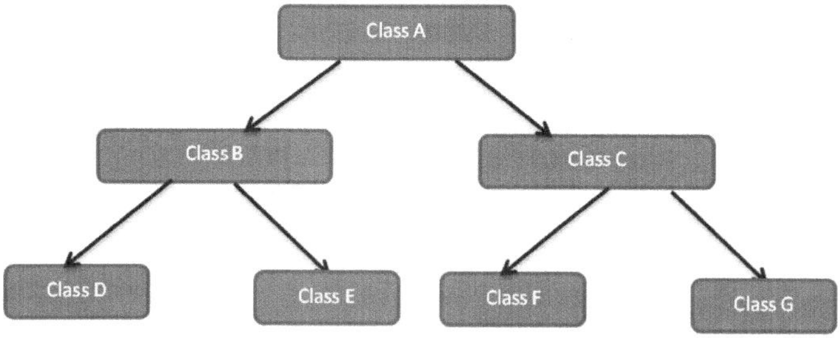


- In the given example, class A has two child class B and class D. Further, class B and class C both are having two child - class D and E; class F and G respectively.
- Syntax for Hierarchical Inheritance

```
//Base Class
class A
{
  public void foo1()
  {
        //to do
  }
}
//Derived Class
class B : A
{
  public void foo2()
  {
        //to do
  }
}
```

ISBN: 978-93-91373-84-9

```
//Derived Class
class C : A
{
  public void foo3()
  {
          //to do
  }
}
//Derived Class
class D : B
{
  public void foo4()
  {
          //to do
  }
}
//Derived Class
class E : B
{
  public void foo5()
  {
          //to do
  }
}
//Derived Class
class F : B
{
  public void foo6()
  {
          //to do
  }
}
```

```
//Derived Class
class G: B
{
  public void foo7()
  {
        //to do
  }
}
```

### 5.2.6. Hybrid inheritance

- This is combination of more than one inheritance. Hence, it may be a combination of Multilevel and Multiple inheritance or Hierarchical and Multilevel inheritance or Hierarchical and Multipath inheritance or Hierarchical, Multilevel and Multiple inheritance.

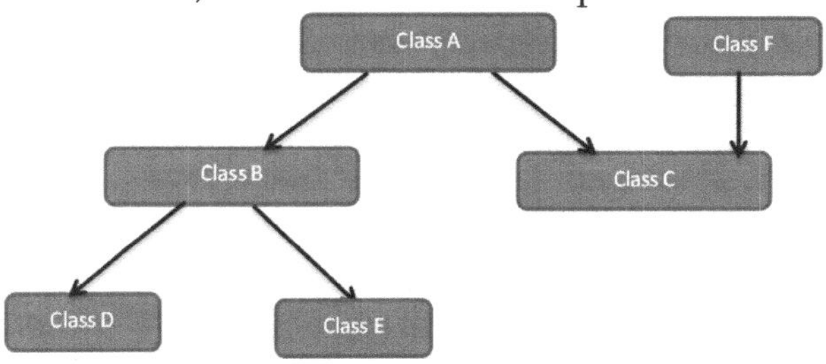


- Syntax for Hybrid Inheritance

```
//Base Class
class A
{
  public void foo1()
  {
        //to do
  }
}
//Derived Class
class B : A
```

ISBN: 978-93-91373-84-9

```
{
  public void foo2()
  {
          //to do
  }
}
//Derived Class
class C : A,F
{
  public void foo3()
  {
          //to do
  }
}
//Derived Class
class D : B
{
  public void foo4()
  {
          //to do
  }
}
//Derived Class
class E : B
{
  public void foo5()
  {
          //to do
  }
}
```

ISBN: 978-93-91373-84-9

### 5.3. Advantages of Inheritance

- Reduce code redundancy.
- Provides code reusability.
- Reduces source code size and improves code readability.
- The code is easy to manage and divided into parent and child classes.
- Supports code extensibility by overriding the base class functionality within child classes.

### 5.4. Disadvantages of Inheritance

- In Inheritance base class and child class, both are tightly coupled. Hence If you change the code of parent class, it will affect all the child classes.
- In a class hierarchy, many data members remain unused and the memory allocated to them is not utilized. Hence it affects the performance of your program if you have not implemented inheritance correctly.

### 5.5. Multilevel Inheritance

- As mentioned earlier java supports multilevel inheritance.
- The syntax for multilevel inheritance is given below.

```
class super1
{
– – –
– – –
}

class sub1 extends super1
{
```

```
        ---
        ---
        }
        class sub2 extends sub1
        {
        ---
        ---
        }
```

- Example: Program to show multilevel inheritance

```
//Super Class
class data1
{
        int a;
        data1(int x)
        {
        a=x;
        }
}
//Sub Class1
class data2 extends data1
{
        int b;
        data2(int x,int y)
        {
        super(x);
        b=y;
        }
}
//Sub Class2
class data3 extends data2
{
        int c;
```

```
        data3(int x,int y,int z)
        {
        super(x,y);
        c=z;
        }
        void compute()
        {
        System.out.println("The Product is " +a*b*c);
        }
}/
/Main Program
class Mainclass
{
        public static void main(String args[])
        {
                data3 d=new data3(1,2,3);
                d.compute();
        }
}
```

## 5.6. Method Overriding/ Overriding Methods

- Methods defined in the superclass is inherited by the subclass and is used by objects created by the subclass
- In some cases we want objects to respond to the same method but have different behaviour when that method is called. This means that we should override the method defined in the superclass.
- This is possible by defining a method in the subclass that has the same name, same arguments and same return type as a method in the superclass.

- Then when that methods is called, the method defined in the subclass is invoked and executed instead of the one in the superclass. This is known as overriding.
- Example: Program to illustrate method overriding

```
class First
{
    int x;
    First(int x)
    {
      this.x=y;
    }
    void display()
    {
      System.out.println("Super x = "+x);
    }
}
class Second extends First
{
    int y;
    Second(int x, int y)
    {
      super(x);
      this.y=y;
    }
    void display()
    {
      System.out.println("Super x = "+x);
      System.out.println("Sub y = "+y);
    }
}
class override
{
```

ISBN: 978-93-91373-84-9

```
    public static void main(String a[])
    {
      Second obj=new Second(23,45);
      First obj1=new First(34);
      obj.display();
      obj1.display();
    }
}
```

**Output:**

```
Super x = 23
Sub y = 45
Super x = 34
```

- When the overridden method display() is invoked for an object of the Second class, the new overridden version of the method is called and not the old definition from the superclass First.

## 5.7. Abstract Methods and Classes

- Abstract methods are those that will have only method declaration without any implementation.
- Hence we can say abstract methods are incomplete methods. A method in java can be declared as abstract by prefixing its declaration with the keyword abstract.
- For example
    - abstract void area();
- The implementations for these abstract methods are provided by overriding the abstract method using a subclass.
- Abstract class is a class that is declared as abstract. An abstract class will have one or more abstract methods.

Abstract classes cannot be instantiated but it can be inherited by other subclasses.

- These subclasses will provide the implementation of all the abstract methods present in the parent class however, if not, the corresponding subclass must be declared abstract.
- Here is an example of a abstract class and its corresponding subclas
- The Coding listed below makes use of an abstract class.

```
abstract class Shape
{
        int l;
        Shape()
        {
                l=10;
        }
        abstract void area(); //abstract method declaration
}
class Rect extends Shape
{
        int b;
        Rect()
        {
                b=10;
        }
        void area
        {
                System.out.println(" Area of Rect is " +l*b);
        }
}
class abs
```

ISBN: 978-93-91373-84-9

```
{
        public static void main(String args[])
        {
                Rect r=new Rect();
                r.area();
        }
}
```

**5.8. Final keyword**

- Java supports an important keyword called Final. The keyword Final is used in three different context.
- Final keyword can be used along with
  - Variables
  - Methods
  - Classes
- When Final keyword is used with variables, it becomes a constant. To declare variable as "Final" prefix the keyword final along with the variable declaration
- Example
  - final int pi=3.14
- When the Final keyword is used with methods, the method cannot be overridden by the subclass method. This prevents method overriding .
- Example

```
  final void calc()
  {
          //body of the method
  }
```

- When the final keyword is used with class, the class cannot be subclassed. Hence this prevents inheritance.
- Example

  final class test

```
{
        //body of the class
}
```

### 5.8.1. Finalizer Methods

- The constructor method is used to initialize an object when it is declared. This process is known as initialization.
- Similarly Java supports a concept called finalization, which is just opposite to initialization.
- We know that Java run-time is an automatic garbage collecting system, which means that it automatically frees up the memory resources used by the objects.
- But sometimes the object may hold other non-object resources such as file descriptors, or window system fonts.
- The garbage collector cannot free these resources. In order to free these resources we must use the finalizer method. This is similar to the destructor in C++.
- The syntax of using finalizer method is
  - finalize()
- This method should explicitly define the task to be performed.
- Finalize () method in java has the following properties
  - every class inherits the finalize() method from java.lang.Object
  - the method is called by the garbage collector when it determines no more references to the object exist
  - the Object finalize method performs no actions but it may be overridden by any class
  - normally it should be overridden to clean-up non-Java resources ie closing a file

ISBN: 978-93-91373-84-9

# CHAPTER 6
# Arrays, Vectors and Strings

## 6. Arrays:

- An array is a group of contiguous or related data items that share a common name.
- The complete set of values is referred to as an array, the individual values are called elements. Array can be of any variable type.
- Arrays are data structures used to hold values that are of the same type. Arrays are fixed-length structures for storing multiple values of the same type.
- Eg: a[100];
- The structure of java array is shown below

a: 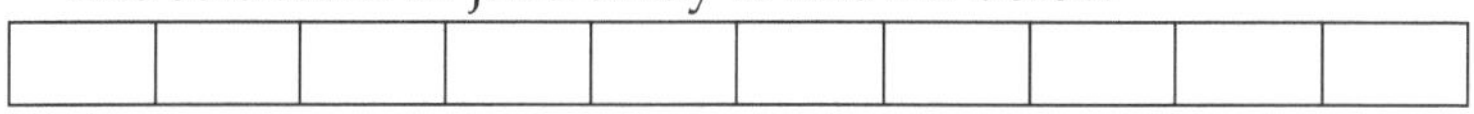

a[0]

- The array 'a' shown in the diagram has ten elements. Like C, C++ the index of an array in java starts with zero.
- All the elements in an array as accessed with the help of array index. The notation a[i] refers to the i-th element of the array.

## 6.1. One-dimensional Arrays:

- A list of items can be given one variable name using only one subscript and such a variable is called a single-subscripted variable or a one-dimensional array.
- In Java, single-subscripted variable $x_1$ can be expressed as
  - X[1], x[2], x[3] .................. x[n]
- The subscript begin with the number 0 ie x[0]

ISBN: 978-93-91373-84-9

### 6.1.1. Java Array Declaration

- An array variable is declared in the same way that any Java variable is declared.It has a data type and a valid Java identifier.
- The data type is the type of the elements contained in the array. The [] notation is used to denote that the variable is an array.
- Java has a unique way of declaring array.
  1. int data[];
     data=new int[50];
  2. int[] data;
     data=new int[50];
  3. int[] data=new int[50];
  4. int data[]=new int[50];
- All the four statements declare one dimensional array called data with the size 50.
- Similarly we can declare array for various data types. Some examples
  - String args[]
  - double points[]

### 6.1.2. Java Array Initialization

- Once an array is declared we can initialize it with some initial values. The syntax is
  - type arrayname[]={list of values };
- For example
  - int data[] = { 5,10,15,20,25 };
- When we initialize an array, we will not specify the size of the array. The java compiler automatically calculates the size the array.
- Example: The following program computes the sum and average of an array

ISBN: 978-93-91373-84-9

```
class Array
{
        public static void main(String args[])
        {
                int data[]= { 5,10,15,20,25 };
                int sum=0;
                double avg;
                for(int i=0;i<data.length;i++)
                sum+=data[i];
                avg=sum/ data.length;
                System.out.println(" The Sum of the Array is
                "+sum);
                System.out.println(" The Average of the Array
                is "+avg);
        }
}
```

**Output of the Program**

The Sum of the Array is 75

The Average of the Array is 15

- **Note:** The size of the array can be found using the variable "length". In the previous example data.length returns the length of the array data.
- **Example:** A simple bubble sort algorithm using single dimensional array is illustrated below.

```
class Bubble
{
        public static void main(String args[])
        {
                int a[]= { 23,7,19,46,53,21 };
                for(int i=0;i<6;i++)
                for(j=i+1;j<6;j++)
```

ISBN: 978-93-91373-84-9

```
            {
                        if(a[i]>a[j])
                        {
                        temp=a[i];
                        a[i]=a[j];
                        a[j]=temp;
                        }
            }
            System.out.println ("The Sorted Output is");
            for(int i=0;i<6;i++)
            System.out.println(a[i]);
      }
}
```

**Output of the Program**

7 19 21 23 46 53

**6.2. Two Dimensional Arrays**

- Often data comes in a two dimensional form. For example, data represented in tables are two dimensional (or more), a computer-generated image is two dimensional, and so on. To handle these data we need 2-D arrays.
- To declare an 2-D array in java, the syntax is
    1. int data[][];
       data=new int[3][3];
    2. int data[][]=new int[3][3]
- The above statement declares a 3x3 matrix. The first dimension represents the row size the second dimension represents the column size.
- In Java, a two-dimensional array x is really an array of one-dimensional arrays:

ISBN: 978-93-91373-84-9

  - int[][] x = new int[3][5];
- The expression x[i] selects the $i^{th}$ one-dimensional array; the expression x[i][j] selects the $j^{th}$ element from that array.

### 6.2.1. Initializing a Two Dimensional Arrays

- We can initialize an 2-D array as given below.
  - int[][] twoD = { {1,2,3}, {4,5,6}, {7,8,9} };
- You can initialize the row dimension without initializing the columns but not vice versa
  - int[][] myArray = new int[5][]; (Correct)
  - int[][] myArray = new int[][5]; (Error will not work)
- Example: The below program performs transpose of a 3x3 matrix.

```
class Transpose
{
	public static void main(String args[])
	{
		int[][] a = { {1,2,3}, {4,5,6}, {7,8,9} };
		int [][] b=new int[3][3];
		for (int i=0;i<3;i++)
		for (int j=0;j<3;j++)
		{
			b[i][j]=a[j][i];
		}
		System.out.println("Transpose Of the Matrix is
		");
		for (int i=0;i<3;i++)
		{
			for (int j=0;j<3;j++)
			{
```

ISBN: 978-93-91373-84-9

```
                    System.out.print(b[i][j]+"\t");
                }
                System.out.println();
            }
        }
}
```

**Output of the Program**

Transpose of the Matrix is

```
1  4  7
2  5  8
3  6  9
```

**6.3. Vectors**

- Vector is a class available in the java.util package. Vectors are commonly used instead of arrays, because they expand automatically when new data is added to them.
- Vectors can hold only Objects and not primitive types (eg, int). If you want to put a primitive type in a Vector, you have to convert the primitive type to object and then add it to the vector.
- You must import either
  - import java.util.Vector; or
  - import java.util.*; to work with the vector.
- Vectors are implemented with an array, and when that array is full and an additional element is added, a new (bigger) array must be allocated and copy the elements from the old array to the new array, if you knew the final size, it is better to use an array but however if you want to create an array that dynamically grow and shrink in size,Vector is a better choice.

ISBN: 978-93-91373-84-9

### 6.3.1. Vector Creation

- To create a Vector Java supports two types of constructors. The first constructor creates a vector with a default size 10.
- The general form is
  - Vector v=new Vector();
- Here v is the object of type vector.
- The second form of constructor creates a vector with a specified size.
- The general form is
  - Vector v=new Vector(30);
- The constructor given above creates a vector with an initial size 30.

### 6.3.2. Methods in Vector

- Vector class in java supports many methods. Some important methods availablein Vector class are tabulated below

| Method | Description |
|---|---|
| v.add(o) | adds Object o to Vector v |
| v.add(i, o) | Inserts Object o at index i, shifting elements up as necessary. |
| v.clear() | removes all elements from Vector v |
| v.contains(o) | Returns true if Vector v contains Object o |
| v.firstElement() | Returns the first element. |
| v.get(i) | Returns the object at int index i. |
| v.lastElement() | Returns the last element. |
| v.listIterator() | Returns a ListIterator that can be used to go over the Vector. This is a useful alternative to the for loop. |

ISBN: 978-93-91373-84-9

| | |
|---|---|
| v.remove(i) | Removes the element at position i, and shifts all following elements down. |
| v.set(i,o) | Sets the element at index i to o. |
| v.size() | Returns the number of elements in Vector v. |
| v.toArray(Object[]) | The array parameter can be any Object subclass (eg, String). This returns the vector values in that array (or a larger array if necessary). This is useful when you need the generality of a Vector for input, but need the speed of arrays when processing the data. |

- Example:This example program makes use of some commonly used vector methods

```
import java.util.*;
class VectorDemo
{
	public static void main(String args[])
	{
		Vector v=new Vector();
		v.add(new Integer(10));
		v.add(new Double(55.10));
		v.add(new String("Vector"));
		v.add(new Integer(100));
		System.out.println(" The First Element is "+v.firstElement());
		System.out.println(" The Last Element is "+v.lastElement());
		System.out.println(" The Size of the Vector is "+v.size());
		System.out.println(" Vector Elements is ");
		ListIterator i=v.listIterator();
```

ISBN: 978-93-91373-84-9 

```
                while(i.hasNext())
                {
                        System.out.println("The Element is
                "+i.next());
                }
                System.out.println("Removing Second Element
                "+ v.remove(2));
                System.out.println("Adding a New Element in
                Third Position");
                v.add(new String("JAVA"));
                System.out.println("The New Set of
                Elements");
                i=v.listIterator();
                while(i.hasNext())
                {
                        System.out.println("The Element is
                "+i.next());
                }
        }
}
```

**Output of the Program**

```
The First Element is 10
The Last Element is 100
The Size of the Vector is 4
Vector Elements is
The Element is 10
The Element is 55.1
The Element is Vector
The Element is 100
Removing Second Element Vector
Adding a New Element in Third Position
The New Set of Elements
The Element is 10
```

ISBN: 978-93-91373-84-9

The Element is 55.1
The Element is 100
The Element is JAVA

## 6.4. Wrapper Classes

- Vectors cannot handle primitive data types like int, float, long, char and double.
- Primitive data types must be converted into object types by using the wrapper classes contained in java.lang package.

| Simple type | Wrapper class |
|---|---|
| Boolean | Boolean |
| Char | Character |
| Double | Double |
| Float | Float |
| Int | Integer |
| Long | Long |

- The wrapper classes have a number of unique methods for handling primitive data types and objects

### 6.4.1. Converting Primitive Numbers to Object Numbers using Constructor Methods

| Constructor calling | Conversion Action |
|---|---|
| Integer Intval=new Integer(i) | Primitive integer to Integer Object |
| Float Floatval=new Float(f) | Primitive float to Float Object |
| Double doubleval=new Double(d) | Primitive double to Double Object |
| Long longval=new Long(l) | Primitive long to Long Object |

ISBN: 978-93-91373-84-9

### 6.4.2. Converting Object Numbers to Primitive Numbers using typeValue() method

| Method calling | Conversion action |
|---|---|
| int i=IntVal.intValue() | Object to primitive integer |
| float f=FloatVal.floatValue() | Object to primitive float |
| long l=longval.longValue() | Object to primitive long |
| double d=doubleval.doubleValue() | Object to primitive double |

### 6.4.3. Converting Numbers to Strings Using toString() Method

| Method calling | Conversion action |
|---|---|
| str=Integer.toString(i) | Primitive integer to string |
| str=Float.toString(f) | Primitive float to string |
| str=Double.toString(d) | Primitive double to string |
| str=Long.toString(l) | Primitive long to string |

### 6.4.4. Converting String Objects to Numeric Objects using the Static Method Valueof()

| Method calling | Conversion action |
|---|---|
| doubleval=Double.Valueof(str) | Converts string to Double object |
| Floatval=Float.Valueof(str) | Converts string to Float object |
| Intval=Integer.Valueof(str) | Converts string to Integer object |
| longval=Long.Valueof(str) | Converts string to Long object |

ISBN: 978-93-91373-84-9

### 6.4.5. Converting Numeric Strings to Primitive Numbers Using Parsing Methods

| Method calling | Conversion action |
|---|---|
| int i=Integer.parseInt(str) | Converts string to primitive integer |
| long i=Long.parseLong(str) | Converts string to primitive long |

- Example: Program to illustrate the use of most commonly used wrapper class methods

```
import java.io.*;
class invest
{
    public static void main(String args[])
    {
      Float p=new Float(0);
      Float r=new Float(0);
      int yrs=0;
      try
      {
        DataInputStream in = new
DataInputStream(System.in);
        System.out.println("Enter the Principal Amt: ");
        System.out.flush();
        String pr=in.readLine();
        p=Float.valueOf(pr);
        System.out.println("Enter the rate of Interest: ");
        System.out.flush();
        String r1=in.readLine();
        r=Float.valueOf(r1);
        System.out.println("Enter the number of years: ");
        System.out.flush();
        String yr=in.readLine();
```

ISBN: 978-93-91373-84-9

```
        yrs=Integer.parseInt(yr);
      }
      catch(IOException e)
      {
        System.out.println("I/O error");
        System.exit(1);
      }
      float value=loan(p.floatValue(),r.floatValue(), yrs);
      printline();
      System.out.println("Final value = "+value);
      printline();
    }
    static float loan(float p, float r, int n)
    {
      int yr=1;
      float sum=p;
      while(yr<=n)
      {
        sum=sum*(1+r);
        yr=yr+1;
      }
      return sum;
    }
    static void printline()
    {
      for(int i=1;i<=30;i++)
      {
        System.out.print("=");
      }
      System.out.println(" ");
    }
}
```

ISBN: 978-93-91373-84-9

**Output**

```
Enter the Principal Amt: 10000
Enter the rate of Interest: 2.4
Enter the number of years: 3
==================
Final value = 393040.0
==================
```

## 6.5. Strings

- String manipulation is the most common part of many Java programs. Strings represent a sequence of characters.
- We can represent strings as a character array, but they will not support the range of operations to be performed on strings.
- In java strings are class objects and implemented using two classes
  - String and
  - StringBuffer
- In Java string is not a character array and is not NULL terminated
- Syntax for declaration
  - String stringName;
  - stringName = new String("string");
- Example:
  - String name= new String("Jackson");

### 6.5.1. String Methods

- The String class in java supports a variety of methods to process strings.
- The class String includes methods

- For handling individual characters
- Comparing string
- Searching string
- Extracting substrings

#### 6.5.1.1. Methods for Comparing and Concatenation of Strings

- Strings in java can be compared using the = operator and equals() method. In java
- Strings can be concatenated using "+" operator and concat() method.

#### 6.5.1.2. Methods for Handling Individual Characters

- The Table below list all the methods to handle individual characters

| Method | Purpose |
|---|---|
| charAt(int index) | Returns the character at the specified index. |
| indexOf(int ch) | Returns the index within this string of the first occurrence of the specified character |
| indexOf(int ch, int fromIndex) | Returns the index within this string of the first occurrence of the specified character, starting the search at the specified index. |
| lastIndexOf(int ch) | Returns the index within this string of the last occurrence of the specified character. |
| lastIndexOf(int chint fromIndex) | Returns the index within this string of the last occurrence of the specified character, searching backward starting at the specified index. |

ISBN: 978-93-91373-84-9

| | |
|---|---|
| replace(char oldChar,char newChar) | Returns a new string resulting from replacing all occurrences of oldChar in this string with newChar |

### 6.5.1.3. Methods for Searching Strings

| Method | Purpose |
|---|---|
| startsWith(String prefix) | Tests if this string starts with the specified prefix. |
| startsWith(String prefix, int toffset) | Tests if this string starts with the specified prefix beginning a specified index |
| endsWith(String suffix) | Tests if this string ends with the specified suffix |

### 6.5.1.4. Methods for Extracting Substrings

| Method | Purpose |
|---|---|
| substring(int beginIndex) | Returns a new string that is a substring of this string. |
| substring(int beginIndex,int endIndex) | Returns a new string that is a substring of this string. |

### 6.5.1.5. Methods for Case Conversion

| Method | Purpose |
|---|---|
| toLowerCase() | Converts all of the characters in this String to lower case |
| toUpperCase() | Converts all of the characters in this String to upper case |

### 6.5.1.6. Miscellaneous methods

| Method | Purpose |
|---|---|
| trim() | This method removes white space from the front and the end of a String. |
| length() | Returns length of the String. |

ISBN: 978-93-91373-84-9

```
class example1
{
    public static void main(String args[])
    {
      String s1="Java";
      String s2="Java";
      String s3="C++";
      String str="Welcome to Java Programming";
      String obj=new String("Java");
      System.out.println("String Manipulation");
      System.out.println("Equal and Concat operations");
      if(s1==s2)
        System.out.println(s1+" and "+ s2+"  are Equal");
      else
        System.out.println(s1+"  and "+ s2+"  are not Equal");
      if(s1.equals(obj))
        System.out.println("Now !! s1 and obj are Equal");
      else
        System.out.println("s1 and obj are not equal");
      System.out.println("s1 + s3 ="+s1+s3);
      System.out.println("s2 + s3 ="+s2.concat(s3));
      System.out.println("Single character handling
methods");
      System.out.println("4th Character "+str.charAt(4));
      System.out.println("Index of Character J is
"+str.indexOf('J'));
      System.out.println("Last Index of Character i is
"+str.lastIndexOf('i'));
      System.out.println("When n Changed to m
"+str.replace('m','n'));
      System.out.println("Searching methods ");
```

ISBN: 978-93-91373-84-9

```
        System.out.println("String Starts with Wel : "+str.startsWith("Wel"));
        System.out.println("Ends With Wel: "+str.endsWith("Wel"));
        System.out.println("Substring from 6th position : "+str.substring(5));
        System.out.println("Substring from 6th to 10th position : "+str.substring(5,9));
        System.out.println("Covert to uppercase: "+str.toUpperCase());
        System.out.println("Convert to lowercase: "+str.toLowerCase());
    }
}
```

**Output of the Program:**

```
String Manipulation
Equal and Concat operations
Java and Java are Equal
Now !! s1 and obj are Equal
s1+ s3 = JavaC++
s1+ s3 = JavaC++
Single character handling methods
4th Character o
Index of Character J is 11
Last Index of Character i is 24
When n Changed to m Welcome to Java Progranning
Searching methods
String Starts with Wel : true
Ends With Wel : false
Substring from 6th position : me to Java Programming
Substring from 6th to 10th position : me t
```

ISBN: 978-93-91373-84-9

Convert to uppercase: Welcome to java programming
Convert to lowercase: Welcome to java programming

### 6.5.2. StringBuffers

- StringBuffer class is a mutable class unlike the String class which is immutable. Both the size and character string of a StringBuffer can be changed dynamically.
- String buffers are preferred when frequent modification of character strings is involved like appending, inserting, deleting, modifying etc.
- A StringBuffer method involves substantial overhead in terms of computational complexity and memory space hence StringBuffer class must be used only in situations where String class cannot be used. Strings can be obtained from string buffers.

#### 6.5.2.1. Creation of StringBuffer

- A StringBuffer in Java can be created using the following three constructors.
    1. StringBuffer strBuf2 = new StringBuffer(100); //With capacity 100
    2. StringBuffer strBuf3 = new StringBuffer(); //Default Capacity 16
    3. StringBuffer strBuf1 = new StringBuffer("Computer"); // Capacity is 24
- The first constructor creates a StringBuffer that has the storage capacity to store 100 characters.
- The second constructor creates a StringBuffer that has the default storage capacity to store 16 characters.
- The last constructor creates a StringBuffer with a String parameter, it creates a StringBuffer with a

ISBN: 978-93-91373-84-9 

capacity 24 i.e default capacity + 8 (the length of String parameter).

### 6.5.2.2. Methods in StringBuffer

- StringBuffer supports many methods that are available in String Class, apart from that StringBuffer supports some unique methods.

| Method | Purpose |
|---|---|
| capacity() | Returns the current capacity of the String buffer. |
| length() | Returns the length (character count) of this string buffer. |
| toString() | Converts to a string representing the data in this string buffer |
| insert(int offset, char c) | Inserts the string representation of the char argument into the StringBuffer |
| delete(int start, int end) | Removes the characters in a substring of this StringBuffer |
| replace(int start, int end, String str) | Replaces the characters in a substring of this StringBuffer with characters in the specified String. |
| reverse() | The character sequence contained in this string buffer is replaced by the reverse of the sequence. |
| append(String str) | Appends the string to this string buffer. |

- Example: The following illustrates the various StringBuffer methods.

```
public class StringBufferMethods
```

ISBN: 978-93-91373-84-9

```
{
        public static void main(String[] args)
        {
                //Examples of Creation of Strings
                StringBuffer        strBuf1        =        new
        StringBuffer("APPLE");
                StringBuffer strBuf2 = new StringBuffer(100);
        //With capacity 100
                StringBuffer  strBuf3  =  new  StringBuffer();
        //Default Capacity 16
                //Print the Capacity of StringBuffers
                System.out.println("strBuf1      capacity      :
        "+strBuf1.capacity());
                System.out.println("strBuf2      capacity      :
        "+strBuf2.capacity());
                System.out.println("strBuf3      capacity      :
        "+strBuf3.capacity());
                //toString Method
                System.out.println("strBuf1   toString()   is   :
        "+strBuf1.toString());
                //append Method
                strBuf2.append("HandBook");
                System.out.println(" strBuf2 is appended with
        :"+strBuf2.toString());
                //insert Method
                strBuf2.insert(1, 'A');
                System.out.println("strBuf2 when A is inserted
        at 1 :"+strBuf2.toString());
                strBuf2.insert(2, 'n');
                System.out.println("strBuf2 when n is inserted
        at 2 :"+strBuf2.toString());
                //delete Method
```

ISBN: 978-93-91373-84-9

```
            strBuf2.delete(2, 'n');
            System.out.println("strBuf2 when n is deleted
    at 2 :"+strBuf2.toString());
            //reverse Method
            strBuf2.reverse();
            System.out.println("Reversed      strBuf2      :
    "+strBuf2);
}
}
```

**Output of the Program**

```
strBuf1 capacity : 21
strBuf2 capacity : 100
strBuf3 capacity : 16
strBuf1 toString() is : APPLE
strBuf2 is appended with :HandBook
strBuf2 when A is inserted at 1 : HAandBook
strBuf2 when n is inserted at 2 : HAnandBook
strBuf2 when n is deleted at 2 : HA
Reversed strBuf2 : AH
```

ISBN: 978-93-91373-84-9

# Chapter 7
# Interface & Package

## 7. Interfaces

- Java does not support multiple inheritance. That is, classes in Java cannot have more than one superclass.
- Hence a definition like this is not permitted in Java

```
class A extends B extends C
{
        -----------
        -------------
}
```

- Many real-life applications require the use of multiple inheritance and so Java provides an alternate approach known as interfaces to support the concept of multiple inheritance.
- A Java class cannot be a subclass of more than one superclass, it can implement more than one interface enabling us to create classes that build upon other classes without the problems created by multiple inheritance.

## 7.1. Defining an Interface

- An interfaceis similar to classes.
- Like classes, interfaces contain methods and variables but with major difference.
- The difference is that interface defines only abstract methods and final fields.
- The interface does not specify any code to implement these methods and data fields contain only constants.

ISBN: 978-93-91373-84-9

- It is the responsibility of the class that implements the interface to define the code for these methods
- The general syntax

```
interface interfacename
{
        variables declaration;
        methods declaration;
}
```

- Here is an example of an interface declaration called clock.

```
interface clock
{
        public int cur_time;
        void getCurrentTime();
        void setCurrentTime();
        void showClock();
}
```

### 7.1.1. Extending Interfaces

- Like classes interfaces can also be extended.
- The new subinterface will inherit all the members of the superinterface.
- Syntax

```
interface name2 extends name1
{
        Body of name2;
}
```

- Example

```
        interface item
        {
                int code = 1001;
                String name="fan";
        }
```

```
interface item1 extends item
{
        void display();
}
```

- The interface item1 inherits the constant code and name into it.
- We can combine several interfaces together into a single interface
- Example

```
interface item
{
    int code = 1001;
    String name="fan";
}
interface methods
{
    void display();
}
Interface item1 extends item, methods
{
        -------------
        ------------
}
```

- While interfaces are allowed to extend to other interfaces, subinterfaces cannot define the methods declared in the superinterfaces.
- It is the responsibility of the class that implements the derived interface to define the methods.

ISBN: 978-93-91373-84-9 

### 7.1.2. Implementing Interfaces

- After creating an interface we can create classes that implement the interface. A class that is implementing an interface should have the following properties.
  - A class that implements the interface must override all the methods present in the interface
  - If a class overrides only a partial set of methods in the interface then in that case the class must be declared as abstract
- The general syntax for a class that implements an interface is given below.

```
[accessmode] class classname [extends superclass] implements interface1,interface2...
{
        //body of the class.
}
```

- An example is given below

```
class tick implements clock
{
        public void getCurrentTime()
        {
                //body of the method.
        }
        public void setCurrentTime()
        {
                //body of the method.
        }
        public void showClock()
        {
                //body of the method.
        }
}
```

- Example: A simple example is given below that give complete details about how to create and make use of an interface.

```
interface arithmetics
{
        public int result;
        void add(int x,int y);
        void sub(int x,int y);
        void mul(int x,int y);
        void div(int x,int y);
}
class calculator implements arithmetics
{
        calculator()
        {
                x=1;
                y=1;
        }
        calculator(int m,int n)
        {
                x=m;
                y=n;
        }
        public void add(int x,int y)
        {
                result=x+y;
                System.out.println("SUM IS "+result);
        }
        public void sub(int x,int y)
        {
                result=x-y;
                System.out.println("DIFFERENCE IS "+result);
```

```
    }
    public void mul(int x,int y)
    {
        result=x*y;
        System.out.println("PRODUCT IS "+result);
    }
    public void div(int x,int y)
    {
        if(x<y)
            System.out.println("RESULT WILL BE
        A REAL VALUE"):
        else
        {
            result=x/y;
            System.out.println("QUOTIENT IS
        "+result);
        }
    }
    public static void main(String args[])
    {
        calculator cal =new calculator(25,5);
        cal.add();
        cal.sub();
        cal.mul();
        cal.div();
    }
}
```

- Since java does not support multiple inheritance directly often we want to extend one parent class and implement many interfaces.
- For example

ISBN: 978-93-91373-84-9

```
class myclass extends Frame implements
ActionListener,ItemListener
{
    // body of the class.
}
```

- Here we can note that the class "myclass" inherits the properties of the parent class Frame and implements to interfaces namely ActionListener and ItemListener.

## 7.2. Accessing Interface Variables

- An interface can contain variables. All the variables that are declared in the interface are final variables.
- Declaring some variables in an interface is similar to declaring #defined constants in C/C++ languages or const variables.
- When you create some variables in an interface without any methods, then the class implementing that interface will not have any method to override
- Hence the class will import only the constants from the interface. The final variables in such interface can act as shared constants.
- Example: The example given below show the procedure to access interface variables.

```
interface constants
{
        int BAD=0;
        int GOOD=1;
}
class test implements constants
{
        int mark;
        ex1(int a)
```

ISBN: 978-93-91373-84-9

```
        {
                mark=a;
        }
        int check()
        {
                if(mark>=50)
                        return GOOD;
                else
                        return BAD;
        }
        public static void main(String args[])
        {
                test e=new test(77);
                int x=e.check();
                if(x==GOOD)
                        System.out.println("Well Done !!!!");
                else
                        System.out.println("Better Luck Next
                Time !!!!");
        }
}
```

### 7.3. Packages: Putting Classes Together

- Code reuse is an important concept in object-oriented programming. One tool that can help us increase the reusability of our code is the package statement.
- Programmers bundle groups of related types into packages.
- A Java package is a mechanism for organizing java classes and interfaces. Packages are also called as containers of classes.

ISBN: 978-93-91373-84-9

- Java packages can be stored in compressed files called JAR files, allowing classes to download faster as a group rather than one at a time.
- Programmers also typically use packages to organize classes belonging to the same category or providing similar functionality. A package in java can contain the following members
  - Classes
  - Interfaces
  - Enumerations
  - Annotations
- By organizing the classes into packages we achieve the following benefits
  - The classes contained in the packages of other programs can be easily reused.
  - In packages, classes can be unique compared with classes in other packages, i.e., two classes in two different packages can have the same name. They can be referred by the package name and the class name
  - Packages provide a way to hide classes thus preventing other programs or packages from accessing classes that are meant for internal use only
  - Packages provide a way for separating design from coding. First the classes can be designed and then their relationship can be decided.
- Java language supports two types of packages. They are
  - Built in Packages (or) Java API packages
  - User Defined Packages

ISBN: 978-93-91373-84-9

## 7.4. Java API Packages

- Java language has a rich set of built in API packages from which you can access interfaces and classes, and then fields, constructors, and methods. The Figure gives the list of Java API packages.

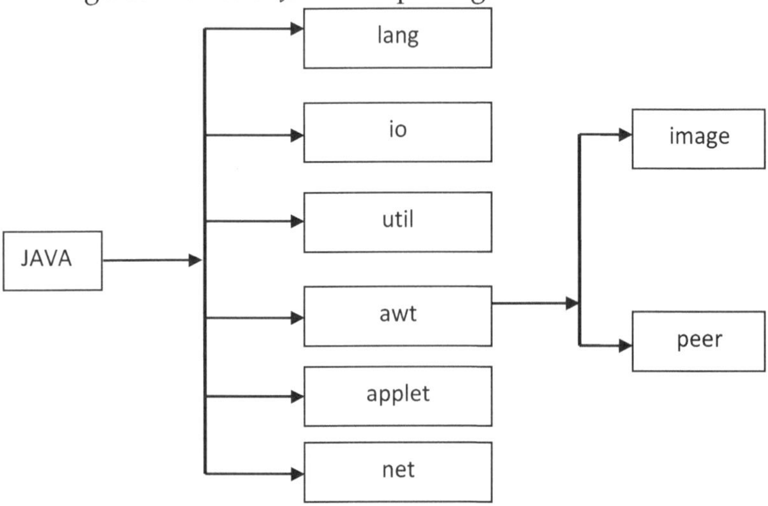


- The details regarding these packages are given below.

| Package name | Contents |
|---|---|
| java.lang | Package that contains essential Java classes, including numerics, strings, objects, compiler, runtime, security, and threads.<br>This is the only package that is automatically imported into every Java program. |
| java.io | Package that provides classes to manage input and output streams to read data from and write data to files, strings, and other sources. |

ISBN: 978-93-91373-84-9

| | |
|---|---|
| java.util | Package that contains miscellaneous utility classes, including generic data structures, bit sets, time, date, string manipulation, random number generation, system properties, notification, and enumeration of data structures. |
| java.net | Package that provides classes for network support, including URLs, TCP sockets, UDP sockets, IP addresses, and a binary-to-text converter. |
| java.awt | Package that provides an integrated set of classes to manage user interface components such as windows, dialog boxes, buttons, checkboxes, lists, menus, scrollbars, and text fields. (AWT = Abstract Window Toolkit) |
| java.awt.image | Package that provides classes for managing image data, including color models, cropping, color filtering, setting pixel values, and grabbing snapshots. |
| java.awt.peer | Package that connects AWT components to their platform-specific implementations (such as Motif widgets or Microsoft Windows controls). |
| java.applet | Package that enables the creation of applets through the Applet class. It also provides several interfaces that connect an applet to its document and to resources for playing audio. |

ISBN: 978-93-91373-84-9

### 7.4.1. Using System Packages/Using Java API Packages

- In Java packages are organized in a hierarchical structure. The Java Package name consists of words separated by periods.
- The first part of the name contains the name java. The remaining words of the Java Package name reflect the contents of the package. The Java Package name also reflects its directory structure.
- For example the statement
  - java.awt -- represents the awt (Abstract Window Toolkit) package. As mentioned earlier the first part of the package name starts with the word Java. The second part of the package name "awt" stands for the contents of the package, in this case the Abstract Window Toolkit.
- To import a class from the package into our program you have to use the import statement. The syntax is
  - import packagename.classname

    or
  - import packagename.*;
- The first form of the syntax imports the specific class from the package. For example the statement import java.util.Date imports the Date class available in the java.util package into our program.
- The second form of the syntax import all the classes available in the specified package. For example java.io.* will import all the classes available in the io package.
- It is always good to import only the classes you need. Bringing in an entire package does not alter the size of the class file, but does lengthen the compile time.

ISBN: 978-93-91373-84-9

### 7.4.2. Creating Packages / User Defined Packages

- You know that java supports a rich set of built in packages. Apart from these packages Java allows us to create our own user defined packages.
- We can create our own package and use it in our programs. Creation of package in java is easy.
- The various steps involved in creation of package are listed below.
    - The package has to be declared at the beginning of the program. The syntax is
        - package packagename;
    - For example
        - package mypack;
    - The second step is to define classes and interfaces and put it inside the package. A package can contain any number of classes; however one of the classes should have the public access specifier. For example

```
package mypack
public class sample
{
        //body of the class
}
```

    - Add methods and variables to class as required
    - Create a subdirectory with name of the package and put it under the current working directory.
    - Save the class in the subdirectory and compile the program.
- Example 1: A simple program to illustrate the creation of user-defined package is given below.

```
package mypack;
public class sample
```

```
{
        public sample()
        {
                System.out.println("This is from the
        constructor");
        }
        public void test()
        {
                System.out.println("This is from the package
        mypack");
        }
}
```

- Create a directory with the name mypack and put inside the current working directory.
- Save the program with the name "sample.java" put in the sub directory mypack and compile the program.
- Once a package is created, you can import the package and use it in your program.

### 7.4.3. Using a Package

- Example 1: Consider the code listing given below.

```
import mypack.sample;
class Client
{
        public static void main(String args[])
        {
                sample s=new sample();
                s.test();
        }
}
```

- The program Client imports the classes from package mypack. The program "Client.java" should be saved in the directory for which mypack is a subdirectory. After that compile and execute the program

- Example 2: In Java it is possible to access a class in one package from another package.

```
package pack1;
class base
{
        public String str="JAVA";
        protected int i=20;
        public void print_string()
        {
                System.out.println("The String is "+str);
        }
        public void print_int()
        {
                System.out.println("The Integer is "+i);
        }
}
```

- The class base is available in the package pack1. Now let's create another package pack2 which contains a class that access the class base

```
package pack2;
import pack1.*;
class X
{
        base b=new base();
        b.print_string();
        b.print_int();
}
```

- Thus in Java it is possible to access a class in one package from another package.

ISBN: 978-93-91373-84-9

**Output of the Program**

The String is JAVA

The Integer is 20

## 7.5. Visibility and Access Rights in Java

| Access modifier | Public | Protected | Friendly | Private protected | Private |
|---|---|---|---|---|---|
| Same class | Yes | Yes | Yes | Yes | Yes |
| Subclass in same package | Yes | Yes | Yes | Yes | No |
| Other classes in same package | Yes | Yes | Yes | No | No |
| Subclass in other packages | Yes | Yes | No | No | No |
| Non-subclass in other packages | Yes | No | No | No | No |

- While using packages and inheritance in a program, we should be aware of the visibility restrictions imposed by various access protection modifiers. As pointed out earlier, packages act as containers for classes and other packages, and classes act as containers for data and methods. Data members and methods can be declared with the access protection modifiers such as private, protected and public as well as default.

ISBN: 978-93-91373-84-9

## 7.6. Adding a Class to a Package

- To add a class to an existing package, consider the following steps
- Example

```
package p1;
public class A
{
        //body of A
}
```

- The package p1 contains one public class by name A. Suppose we want to add another class B to this package, this can be done as follows
  - Define the class and make it public
  - Place the package statement

    ```
    package p1
    ```

  - Before the class definition as follows

    ```
    package p1;
    public class B
    {
            //body of B
    }
    ```

  - Store this as B.java file under the directory p1
  - Compile B.java. This will create a B.class file and place it in the directory p1.
  - We can also add a non-public class to a package using the same procedure
  - The statement import p1 will import both the classes A and B
- Since java source file can have only one class declared as public, we cannot put two or more public classes together in a .java file. This is because of the restriction

ISBN: 978-93-91373-84-9

that the file name should be same as the name of the public class with .java extension.

- If we want to create a package with multiple public classes in it, we may follow the following steps
  - Decide the name of the package
  - Create a subdirectory with this name under the directory where main source files are stored
  - Create classes that are to be placed in the package in separate source files and declare the package statementpackage packagename;
  - Switch to the subdirectory created earlier and compile each source file. When completed, the package would contain .class files of all the source files.

## 7.7. Hiding Classes

- When a package is imported to your program all the public classes will be imported to your program.
- Sometimes it may be desired to hide some classes in the package because accessing them may lead to some problems. To hide such classes these classes must be declared with a access specifier other than public.
- These non public classes will be hidden when the package is imported to another program. Consider the example

```
package pack1;
public class Publicclass
{
//body of the class
}
class Defaultclass
{
```

ISBN: 978-93-91373-84-9 

```
//body of the class
}
```

- In this example the class Publicclass with the public scope is visible to programs importing the package pack1.
- The class Defaultclass with no access specifier is hidden to programs importing the package pack1. Thus,

```
import pack1.*;
class test
{
        public static void main(String args[])
        {
                Publicclass p1; //OK
                Defaultclass d1; //ERROR
                .....
                .....
        }
}
```

ISBN: 978-93-91373-84-9 

# CHAPTER 8
# Threads in Java

## 8.1. Multithreaded Programming

- The earliest computers did only one job at a time. All programs were run sequentially, one at a time; and each had full run of the computer.
- However two programs couldn't be run at once. This kind of processing is called batch processing.
- It is one of the earliest computing technologies
- Time sharing systems were invented after batch processing to allow multiple people to use the computer at the same time. On a time sharing system many people can run programs at the same time. The operating system is responsible for splitting the time among the different programs that are running.
- Once systems allowed different users to run programs at the same time the next step is to allow the same user run multiple programs simultaneously.
- Each running program (generally called a process) had its own memory space, its own set of variables, its own stack and heap, and so on. The ability to execute more than one task at the same time is called Multitasking.
- The terms multitasking and multiprocessing are often used interchangeably, although multiprocessing sometimes implies that more than one CPU is involved.
- In multitasking, only one CPU is involved, but it switches from one program to another so quickly that it gives the appearance of executing all of the programs at the same time.

ISBN: 978-93-91373-84-9

- The system's terminology of multitasking is multithreading. In multithreading a process is divided into two or more sub processes, these sub processes are allowed to execute simultaneously.
- For example a program, can print a file at the same time it can downloads a page and formats the page as it downloads.
- Threads are also called as lightweight processes.

## 8.2. Threads in Java

- Thread is a sequential flow of control. In Java every program consists of at least one thread - the one that runs the main method of the class provided as a startup argument to the Java virtual machine ("JVM").
- Thread in java has a beginning part, the body part and the termination part. All the programs discussed in the previous chapters are single threaded programs however Java supports multithreading and hence in Java it is possible to create multiple flow of control and allow them to execute simultaneously.
- A program that has multiple flow of control is called Multithreading.
- Multithreading is one of the most powerful feature available in java. As mentioned earlier every java program has a single default thread (the one that runs the main method) that controls the execution of the code.
- This main() method thread is the first thread that will start its execution. We can use this main method thread to create many child threads to implement multithreaded programming.

ISBN: 978-93-91373-84-9 

- All the child threads created by the main thread will complete its execution part before the main thread. Thus in a java program main() method thread is the first thread to start and it is the last one to finish its execution.
- In a multithreaded program all the threads executes concurrently and share resources.
- Since one processor is responsible for executing all the threads, the java interpreter performs switching of control between the running threads. This concept is known as context switching in a thread.

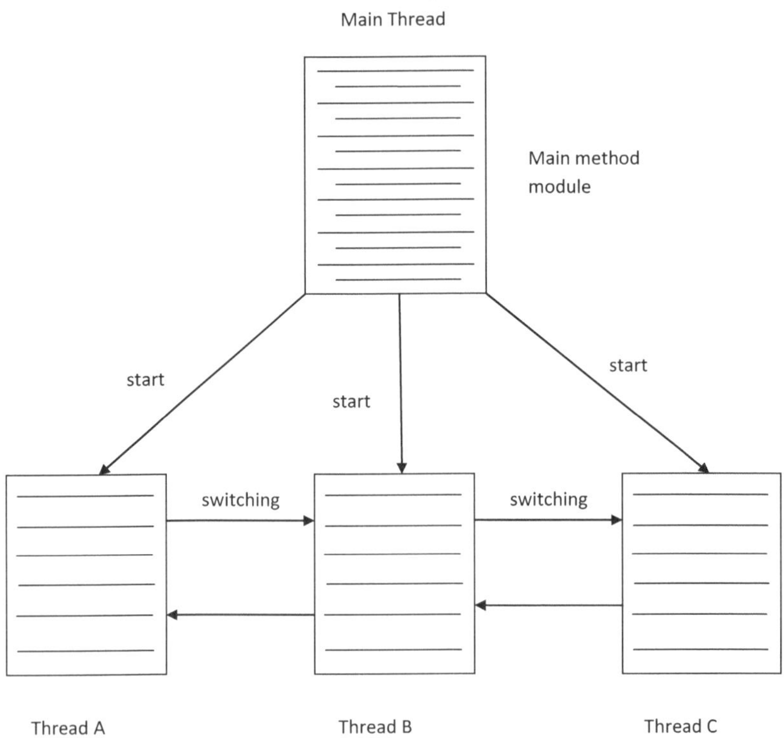


ISBN: 978-93-91373-84-9

## 8.3. Creating Threads

- To create threads, Java has a class called Thread and this class is available in the java.lang package.
- Java supports two different ways of creating threads. They are
  - Extending the Thread class
  - Implementing the Runnable interface
- In both the cases we have to override the method called run(). The run() method will contain the code required by the thread.
- The syntax of run() method is given below.

```
public void run()
{
        // body of method
}
```

### 8.3.1. Extending Thread class

- The simple way to create a threaded program is to extend the Thread class.
- Any class that extends Thread class can access all the methods available in the thread.
- The steps to create a threaded program are given below.
- Step1: Create a class that extends thread. For example

```
class myclass extends Thread
{
        // Body of the class
}
```

- Step 2: Override the run() method and write the code required by the thread

```
class myclass extends Thread
{
        ......
```

ISBN: 978-93-91373-84-9

```
        ......
        public void run()
        {
                //body of the run method.
        }
        ......
        ......
    }
```

- Step 3: Create the object for the threaded class and start the thread using start() method.
- For example
  myclass mc=new myclass()
  mc.start()
- Example 1: The following is a simple example for multithreaded program

```
class Two_table extends Thread
{
    public void run()
    {
        for(int i=1;i<=10;i++)
        {
            System.out.println(i + " * 2 = " + i*2)
        }
    }
}
class Three_table extends Thread
{
    public void run()
    {
        for(int i=1;i<=10;i++)
        {
            System.out.println(i + " * 3 = " + i*3)
```

ISBN: 978-93-91373-84-9

```
                }
        }
}
class mainprg
{
        public static void main(String args[])
        {
                Two_table two=new Two_table();
                Two.start();
                Three_table three=new Three_table();
                Three.start();
                Four_table four=new Four_table();
                four.start();
        }
}
```

- Output of the program.

```
1 * 2 = 2
1 * 3 = 3
2 * 3 = 6
3 * 3 = 9
4 * 3 = 12
5 * 3 = 15
6 * 3 = 18
7 * 3 = 21
8 * 3 = 24
9 * 3 = 27
10 * 3 = 30
2 * 2 = 4
3 * 2 = 6
4 * 2 = 8
5 * 2 = 10
6 * 2 = 12
```

ISBN: 978-93-91373-84-9

7 * 2 = 14
8 * 2 = 16
9 * 2 = 18
10 * 2 = 20

### 8.3.2. Implementing Runnable Interface

- The next method of creating a thread is to create a class and making that class to implement runnable interface.
- Since java does not support multiple inheritance directly we often use runnable interface for creating threads.
- The various steps for creating a multithreaded program in this method are listed below.
- Step1: Create a class that implements runnable interface. For example

```
class myclass implements Runnable
{
// Body of the class
}
```

- Step 2: Override the run() method and write the code required by the thread

```
class myclass implements Runnable
{
        ......
        ......
        public void run()
        {
                //body of the run method.
        }
        ......
        ......
}
```

ISBN: 978-93-91373-84-9

- Step 3: Create the object for the Thread and pass the object of the class "myclass" as a parameter.
- For example
  myclass mc=new myclass()
  Thread t=new Thread(mc);
- Example: The program given in the example1 is modified by implementing the runnable interface.

```
class Two_table implements Runnable
{
        public void run()
        {
                for(int i=1;i<=10;i++)
                {
                        System.out.println(i + " * 2 = " + i*2)
                }
        }
}
class Three_table implements Runnable
{
        public void run()
        {
                for(int i=1;i<=10;i++)
                {
                        System.out.println(i + " * 3 = " + i*3)
                }
        }
}
class mainprg
{
        public static void main(String args[])
        {
                Two_table two=new Two_table();
```

ISBN: 978-93-91373-84-9

```
                Three_table three=new Three_table();
                Thread t1=new Thread(two);
                t1.start();
                Thread t2=new Thread(three);
                t2.start()
        }
}
```

- **Output of the Program**

```
1 * 2 = 2
2 * 2 = 4
3 * 2 = 6
4 * 2 = 8
5 * 2 = 10
6 * 2 = 12
7 * 2 = 14
8 * 2 = 16
9 * 2 = 18
10 * 2 = 20
1 * 3 = 3
2 * 3 = 6
3 * 3 = 9
4 * 3 = 12
5 * 3 = 15
6 * 3 = 18
7 * 3 = 21
8 * 3 = 24
9 * 3 = 27
10 * 3 = 30
```

ISBN: 978-93-91373-84-9 

## 8.4. Stopping and Blocking a Thread

### Stopping a Thread

- Whenever we want to stop a thread from running, call the stop() method.
- Example
  - t1.stop();
- This statement cause the thread to move to the dead state. A thread will also move to the dead state automatically when it reaches the end of its method.
- The stop() method is used when the premature death of a thread is desired.

### Blocking a Thread

- A thread can be temporarily suspended or blocked from entering the running state using either of the following thread methods
  - sleep()
  - suspend()
  - wait()
- These methods cause the thread to go into the blocked or not-runnable state.
- The thread will go to the runnable state when
  - The specified time is elapsed in the case of sleep() method
  - The resume() method is called in case of suspend() method
  - The notify() method is called in case of wait() method

ISBN: 978-93-91373-84-9

## 8.5. Life Cycle of Thread

- When a new thread is created it passes through various states before it gets terminated and this process is called the Life Cycle of the Thread.
- During the life cycle of a thread, there are many states it can enter. They are
  - Newborn state
  - Runnable state
  - Running state
  - Blocked state
  - Dead state
- A thread is always in any one of these states. It can move from one state to another via a variety of ways
- The following figure shows the states a thread can pass in during its life and illustrates which method calls cause a transition to another state.
- Once the Thread class is instantiated a new thread is created and its lifecycle begins
- A newly created thread enters the active running state when the start()method of the thread is invoked
- A running thread may be blocked, when blocked it enters the inactive notrunning state and waits until it is notified to run. A thread may be blockedbecause
  - It has been put to sleep for a set amount of time
  - It is suspended with a call to suspend() and will be blocked until aresume() message
  - The thread is suspended by call to wait(), and will become runnableon notify or notifyAll message.
- Once the thread completes its run() method or when killed by some otherprocess it enters the dead state, that is the end of thread's life cycle.

ISBN: 978-93-91373-84-9 

### 8.5.1. Thread Methods

- Apart from the start() and run() method, a thread class supports the followingmethods that causes the state transition of a thread.
  - sleep() method
  - yield method()
  - stop() method

#### 8.5.1.1. Sleep Method

- The sleep method is static and pauses execution for a set number of milliseconds,
- The sleep method throwsInterruptedException.
- Example: The following program illustrates sleep method of Thread class.

```
public class ThreadSleep extends Thread
{
	public void run()
	{
		try
		{
			for(int i=0;i<5;i++)
			{
				this.sleep(1000);
				System.out.println("looping for
			"+ i +" time");
			}
		}
		catch(InterruptedException ie){}
	}
	public static void main(String args[])
	{
		ThreadSleep t = new ThreadSleep();
		t.start();
```

ISBN: 978-93-91373-84-9

```
        }
}
```

#### 8.5.1.2. Yield Method

- Java Thread class has a static method called yield which causes the currentlyrunning thread to yield its hold on CPU cycles.
- This thread returns to the "readyto run" state and the thread scheduling system has a chance to give other threadsthe attention of the CPU.
- If no other threads are in a "ready to run state" the threadthat was executing may restart running again.
- Example:The following program illustrates yield method of Thread class.

```
class ThreadYeild extends Thread
{
        public void run()
        {
                for(int i=0;i<5;i++)
                {
                        System.out.println("looping for "+ i +"
                time");
                        if(i==4) yield();
                }
        }
}
```

#### 8.5.1.3. Stop Method

- Stop method in java forces the thread to stop executing.
- Example:The following program illustrates stop method of Thread class.

```
class ThreadStop extends Thread
{
        public void run()
        {
                for(int i=0;i<5;i++)
                {
                        System.out.println("looping for
                "+ i +" time");
                        if(i==4) stop();
                }
        }
        public static void main(String args[])
        {
                ThreadStop st=new ThreadStop();
                st.start();
        }
}
```

## 8.6. Thread Priority

- When a Java thread is created, it inherits its priority from the thread that createdit.
- Each Thread has a priority, ranging between Thread.MIN_PRIORITY andThread.MAX_PRIORITY (defined as 1 and 10 respectively).
- The higher the integer,the higher the priority. At any given time, when multiple threads are readyto be executed, the runtime system chooses the "Runnable" thread with the highestpriority for execution. Only when that thread stops, yields, or becomes notrunnable for some reason the lower priority thread start executing.

ISBN: 978-93-91373-84-9

- If two threadsof the same priority are waiting for the CPU, the scheduler chooses one of themto run in a round-robin fashion. Here are some important points regarding threadpriority.
- By default, each new thread has the same priority as the thread thatcreated it. The initial thread associated with a main by default haspriority Thread.NORM_PRIORITY (5).
- The current priority of any thread can be accessed via methodgetPriority().
- The priority of any thread can be dynamically changed via methodsetPriority().
- Example:This program creates threads with different priorities.

```
class A extends Thread
{
	public void run()
	{
		for(int i=1;i<=5;i++)
		{
			System.out.println(i+" From A ")
		}
	}
}
class B extends Thread
{
	public void run()
	{
		for(int i=1;i<=5;i++)
		{
			System.out.println(i+" From B ")
		}
```

ISBN: 978-93-91373-84-9

```
        }
}
class C extends Thread
{
        public void run()
        {
                for(int i=1;i<=5;i++)
                {
                        System.out.println(i+" From C")
                }
        }
}
class mainprg
{
        public static void main(String args[])
        {
                A a=new A();
                A.setPriority(Thread.MIN_PRIORITY);
                a.start();
                B b=new B();
                b.setPriority(b.getPriority()+1);
                b.start();
                C c=new C();
                c.setPriority(Thread.MAX_PRIORITY);
                four.start();
        }
}
```

**8.7. Synchronization**

- Generally in a multithreaded environment all the threads share some critical resourcessuch as a block of code. In many situations these resources has to

ISBN: 978-93-91373-84-9

beexclusively accessed, otherwise strange bugs can arise.

- Java provides a way tolock the shared resources for a thread which is currently executing it, and makingother threads that wish to use it wait until the first thread is finished.
- These otherthreads are placed in the waiting state.
- To lock such shared resources java supports synchronization of threads i.e. providingan exclusive access to a thread currently accessing the shared resource.
- Java has a keyword called "synchronized" that has to be prefixed before the methoddeclaration that contains the code to access the shared resource. For example

```
synchronized void access_resource()
{
        //block of the code to be synchronized.
        ............
        ............
}
```

- When we declare a method as synchronized, Java creates a monitor (which is thebasis of thread synchronization).
- A monitor is an object that can block and revivethreads.
- A monitor is simply a lock that serializes access to an object or a class.
- To gainaccess, a thread first acquires the necessary monitor, and then proceeds. Thishappens automatically every time when a thread enters a synchronized method.
- During the execution of a synchronized method, the thread holds the monitor forthat method's object, or if

ISBN: 978-93-91373-84-9

the method is static, it holds the monitor for that method'sclass.

- If another thread is executing the synchronized method, the current threadis blocked until that thread releases the monitor (by either exiting the method orby calling wait()).

ISBN: 978-93-91373-84-9

# CHAPTER9
# Exception Handling & Applet Programming

## 9. Managing Errors and Exceptions

- Errors are wrongs that can make a program go wrong.
- An error may produce an incorrect output or may terminate the execution of the program abruptly or may even cause the system to crash

## 9.1. Types of Errors

- Errors may broadly be classified into two categories
    - Compile-time errors
    - Run-time errors
- Compile-Time Errors:
    - All syntax errors will be detected and displayed by the Java compiler and therefore these errors are known as compile-time errors.
    - Whenever an error is display the .class file is not created .
    - It is therefore necessary to fix all the errors and then compile the program successfully to execute the program
    - Most of the compile-time errors are due to typing mistakes. The most common errors are
        - Missing semicolons
        - Missing brackets in classes and methods
        - Misspelling of identifiers and keywords
        - Use of undeclared variable
        - Use of = to in place of == operator etc.

ISBN: 978-93-91373-84-9

- Run-Time Errors:
  - Sometimes, the program may compile successfully creating the .class file but may not run properly. These programs may produce wrong results due to wrong logic or may terminate due to errors
  - Most common run-time errors are
    - Dividing an integer by zero
    - Accessing an element that is out of the bounds of an array
    - Trying to store a value into an array of an incompatible class or type
    - Passing a parameter that is not in a valid range or value for a method
    - Trying illegally to change the state of the thread
    - Converting invalid string to a number etc.

## 9.2. Exceptions

- An exception is a condition that is caused by a run-time error in the program.
- When the Java Interpreter encounters an error, it creates an exception object and throws it
- If the exception object is not caught and handled properly, the interpreter will display an error message and terminate the program
- The exception handling mechanism will handle the exception thrown and execute the remaining part of the program without abruptly terminating it.
- The purpose of the exception handling mechanism is to provide a means to detect and report an exceptional circumstance so that appropriate action can be taken.

ISBN: 978-93-91373-84-9

- The exception handling mechanism performs the following task
    - Find the problem (***Hit*** the exception)
    - Inform that an error has occurred (***Throw*** the exception)
    - Receive the error information (***Catch*** the exception)
    - Take corrective action (***Handle*** the exception)
- The error handling code consist of two segments
    - One to detect errors and throw the exception
    - The other to catch and handle the exception
- Some common exceptions are

| **Exception Type** | **Cause of Exception** |
|---|---|
| Arithmetic Exception | Cause by math errors such as divide by zero |
| Array Index Out Of Bounds Exception | Caused by bad array indexes |
| File Not Found Exception | Caused by an attempt to access a nonexistent file |
| IO Exception | Caused by general I/O failures |

### 9.2.1. Syntax of Excepting Handling Code

- The concept of exception handling are throwing an exception and catching it. It is shown below

ISBN: 978-93-91373-84-9

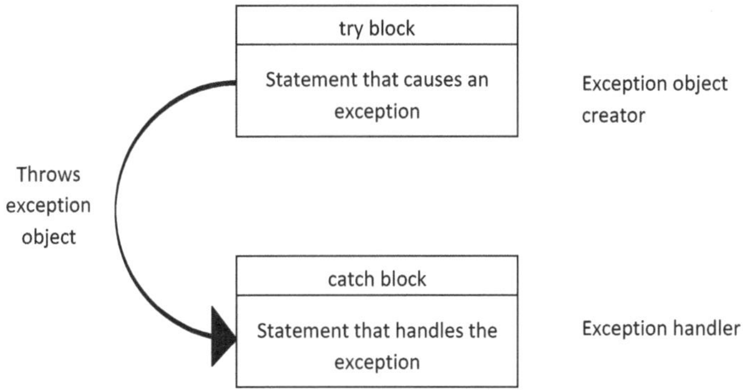


- Java uses the keyword **try** to preface a block of code that is likely to cause an error and throw an exception
- A catch block defined by the keyword **catch** to catch the exception and handle it
- Syntax

```
try
{
        Statement;      //generate an exception
}
catch(Exception-type e)
{
        Statement;      //process the exception
}
```

- The try block should be followed by atleast one catch statement otherwise an error will occur.

### 9.3. Multiple Catch statements

- The try block can have multiple catch statements and in this case the catch statements will work like different cases in the switch structure.
- If an error is thrown it checks the catch statements and the one that first matches the error will be executed.
- Syntax

ISBN: 978-93-91373-84-9

```
    try
    {
            ------------
            -----------
    }
    catch(Exception-Type1 e)
    {
  -----------
  -----------
}
catch(Exception-Type2 e)
{
  -------------
  -------------
}
.
.
.
    catch(Exception-TypeN e)
{
  -------------
  -------------
}
```

## 9.4. Using Finally Statement

- Java supports another statement known as finally that can be used to handle an exception that is not caught by any of the previous catch statement.
- Finally block can be used to handle any exception generated within a try block
- It can be placed after the try block or after all the catch block

ISBN: 978-93-91373-84-9

- Syntax

```
try
{
            ------------
            -----------
}
catch(Exception-Type1 e)
{
   -----------
   -----------
 }
 catch(Exception-Type2 e)
 {
   -------------
   -------------
 }
 .
 .
 .
       catch(Exception-TypeN e)
 {
   -------------
   -------------
 }
          finally
       {
              -----------
              -----------
       }
```

- Example: Program that illustrates the usage of try and catch blocks

```
class Ex
```

ISBN: 978-93-91373-84-9

```
{
    public static void main(String a[])
    {
      int b[] = { 5, 10 };
      int c=5;
      try
      {
        int x=b[2]/c - b[1];
      }
      catch(ArithmeticException e)
      {
        System.out.println("Division by zero");
      }
      catch(ArrayIndexOutOfBoundsException e)
      {
        System.out.println("Array index error");
      }
      finally
      {
        System.out.println("Program generated an error");
      }
      int y=b[1] / b[0];
      System.out.println("Y= "+y);
    }
}
```

### 9.5. Throwing Our Own Exception

- There may be times when we would like to throw our own exceptions. This is done using the keyword throw
- Syntax
  - throw new throwable_subclass;

ISBN: 978-93-91373-84-9

- Exception is a subclass of Throwable and myexception is a subclass of Throwable class
- Example: Programs to show the user defined exception

```
import java.lang.Exception;
class Myexception extends Exception
{
    Myexception(String mess)
    {
      super(mess);
    }
}
class Ex1
{
    public static void main(String a[])
    {
      int x=5, y=1000;
      try
      {
        float z=(float) x/ (float) y;
        if(z<0.01)
        {
              throw new Myexception("Number is too
small");
        }
      }
      catch(Myexception e)
      {
        System.out.println("Caught my exception");
        System.out.println(e.getMessage());
      }
      finally
      {
```

ISBN: 978-93-91373-84-9

```
        System.out.println("Completed the program execution");
      }
    }
}
```

## 9.6. Applet Programming

- An applet is a Java program that runs in a Web browser.
- **Local and Remote Applets**
  - We can embed applets into Web pages in two ways
    - We can write our own applet and embed them into the Web page
    - We can download an applet from a remote computer system and then embed it into the Web page
  - An applet developed locally and stored in a local system is known as local applet
  - A remote applet is that which is developed by someone else and stored on a remote computer connected to the Internet.

## 9.7. How Applets Differ from Applications

- There are certain differences between Applet and Java Standalone Application as described below
  - An applet is a java class that extends the java.applet.Applet class
  - An applet will not define main()
  - Applets are designed to be embedded within an HTML page.

ISBN: 978-93-91373-84-9

- When a user views an HTML page that contains applet, the code for that applet is downloaded to the user's machine.
- Applets have strict security rules that are enforced by the Web browser. The security of an applet is often referred to as sandbox security, comparing the applet to a child playing in a sandbox with various rules that must be followed.

## 9.8. Applet Life Cycle

- Every Java applet inherits a set of default behaviours from the Applet class. The Applet class is a subclass of the Panel class which is again a subclass of the Container class as shown below.

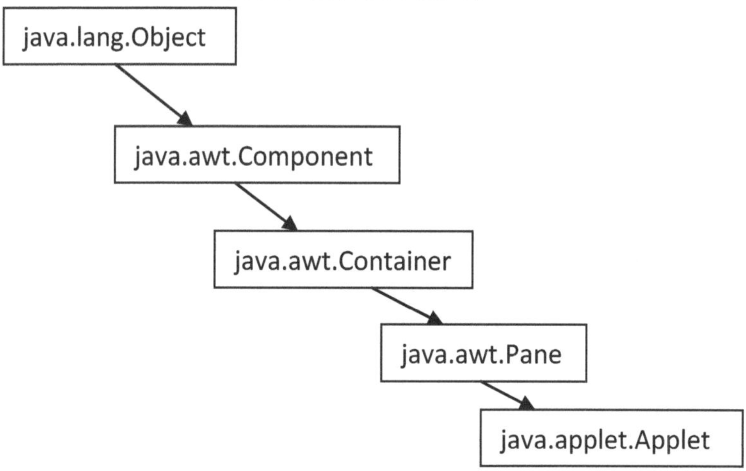


- When an applet is loaded, it undergoes a series of changes in its state. The states include
  - Born or initialization state
  - Running state
  - Idle state
  - Dead or destroyed state

ISBN: 978-93-91373-84-9

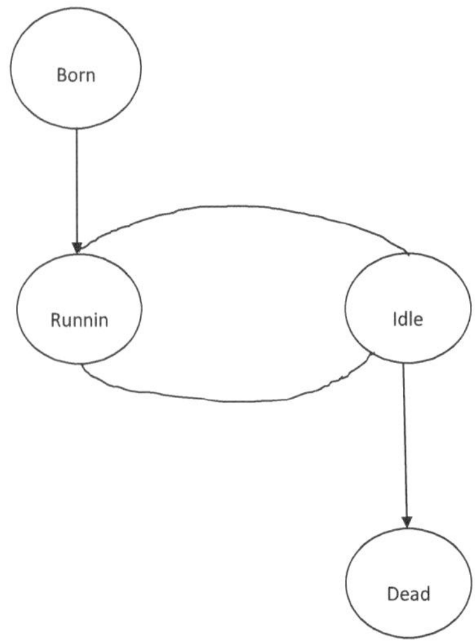


### 9.8.1. Initialization State:

- Applet enters the initialization state when it is first loaded.
- It is achieved by calling the init() method of the Applet class. This method is used for whatever initialization is needed for your applet
- The initialization occurs only once in the applet's life cycle.
- To provide any additional behaviour we must override the init() as follows

```
public void init()
{
            ------------
            ------------
}
```

### 9.8.2. Running state

- Applet enters the running state when the system calls the start() method.

ISBN: 978-93-91373-84-9

- This method is automatically called after the browser calls the init() method. It is also called whenever the user returns to the page containing the applet having gone off to other pages.
- We can override the method start() to create a thread to control the applet

```
public void start()
{
        --------------
        ----------------
}
```

### 9.8.3. Idle or Stopped state

- An applet becomes idle when it is stopped from running.
- Stop occurs automatically when we leave the page containing the applet. This method can be called repeatedly in the same applet.
- The method stop() can be overridden to stop the thread

```
public void stop()
{
        -------------
        -------------
}
```

### 9.8.4. Dead state

- An applet is said to be dead when it is removed from memory. This occurs automatically by invoking the method destroy().
- This method is called when the browser shuts down normally.

ISBN: 978-93-91373-84-9

- We can override the destroy() method to clean up the resources

```
public void destroy()
{
        --------------
        --------------
}
```

### 9.8.5. Display State

- Applet moves to the display state when it has to perform some output operations on the screen.
- The paint() method is called to do this task. This method is actually inherited from the java.awt package.
- We must override this method to display something on the screenpublic void paint(Graphics g)

```
{
        ------------
        ------------
}
```

## 9.9. Creating an Executable Applet

- Executable applet is nothing but the .class file of the applet.
- To create an Applet program follow the steps
    - Create a Java file containing Applet Code and Methods
    - Create a HTML file and embed the .class file of the java file created in the first step
    - Run applet using either of the following methods
        - Open the HTML file in java enabled web browser

ISBN: 978-93-91373-84-9

- Use appletviewer tool

```
//code for MyFirstApplet.java
import java.applet.*;
import java.awt.*;
public class MyFirstApplet extends Applet
{
  public void paint(Graphics g)
  {
    g.drawString("My First Applet Program", 100,100);
  }
}
//Code for MyFirstApplet.html
<html>
<applet code="MyFirstApplet.class" width=300
height=300>
</applet>
```

Embedding the .class file of java code in HTML file

## 9.10. Applet Tag

- The applet tag appears as a pair <applet> .........</applet>
- This tag supplies the name of the applet to be loaded and tells the browser how much space the applet requires.
- The ellipse in the <applet ...> tag indicates that it contains certain attributes that must be specified.
- The applet tag given below specifies the minimum requirements to place the applet on the web page

```
<APPLET
        CODE="hello.class"HEIGHT="300"
        WIDTH="300">
</APPLET>
```

ISBN: 978-93-91373-84-9 

- The above applet tag specifies three things
  - Name of the applet
  - Height of the applet in pixels
  - Width of the applet in pixels
- The applet in its simplest form creates a space of required size and then displays the applet output in that space.
- The syntax of the <APPLET> tag is a little more complex and it includes several attributes.
- Syntax

```
<APPLET
        [CODEBASE = codebase_URL]
        CODE = AppletFileName.class
        [ALT = alternate_text]
        [NAME = applet_instance_name]
        WIDTH = pixels
        HEIGHT = pixels
        [ALIGN = alignment]
        [VSPACE = pixels]
        [HSPACE = pixels]
>
[PARAM NAME= name1 VALUE = value1>]
[PARAM NAME= name2 VALUE = value2>]
.................
.....................
</APPLET>
```

- The attributes inside the [ ] indicate that they are optional and the APPLET TAG requires minimum three attributes

ISBN: 978-93-91373-84-9

| Attribute | Meaning |
| --- | --- |
| CODE = AppletFileName.class | Specifies the name of the applet class to be loaded. This is a compulsory attribute. The class file is already compiled and stored in the same directory |
| CODEBASE= codebase_URL | Specifies the URL of the directory in which the applet resides. |
| WIDTH = pixels<br>HEIGHT = pixels | This attribute specifies the space on the HTML page that will be reserved for the applet |
| NAME = applet_instance_name | A name for the applet is optionally given so that other applets on the page may refer to this applet. This helps in inter-applet communication |
| ALIGN = alignment | This optional attribute specifies where on the page the applet will appear. The possible alignments are TOP, BOTTOM, LEFT, RIGHT, MIDDLE, ABSMIDDLE, ABSBOTTOM, TEXTTOP, BASELINE |
| HSPACE = pixels | This is used only when the ALIGN is set to LEFT or RIGHT. This specifies the amount of horizontal blank space the browser should leave surrounding the applet |
| VSPACE = pixels | This is used only when the ALIGN is set to TOP, BOTTOM, etc. This specifies the amount of vertical blank space the browser should leave surrounding the applet |
| ALT = alternate_text | Non-Java browsers will display this text where the applet is to load. It is an optional attribute. |

## 9.11. Passing Parameters to Applets

- <PARAM ...> tag is used to pass parameter value from HTML file to Applet code.
- Parameters are passed to an applet when it is loaded. We can define the init() method in the applet to get hold of the parameters defined
- This is done using the getParameter() method which takes one string argument representing the name of the parameter and returns a string containing the value of the parameter.
- The following example shows the usage of the parameter tag

```
// code for param_test.html
<html>
<applet code="param_test.class" height="400" width="300">
<PARAM NAME="name" VALUE="Jack">
</applet>
</html>
```

> PARAM name is used as "name" which should be passed as parameter in get Parameter() method to fetch the value of the parameter

```
//code for param_test.java"
import java.applet.*;
import java.awt.*;
public class param_test
extends Applet
{
    Stirng str=null;
    public void init()
    {
      str=getParameter("name");
    }
    public void paint(Graphics g)
    {
```

```
        g.drawString(str,100,100);
    }
}
```

**9.12. Getting Input from User**

- Applets work in Graphics environment where inputs are treated as Strings.
- In order to get input from user, we need to create TextField. We can add this in the HTML coding.
- Example: it shows the input from the user and displays the square of the number

```
import java.applet.*;
import java.awt.*;
public class usr_inp extends Applet
{
    TextField t1,t2;
    public void init()
    {
        t1=new TextField(5);
        t2=new TextField(5);
        add(t1);
        add(t2);
    }
    public void paint(Graphics g)
    {
        int x=0;
        x=Integer.parseInt(t1.getText());
        t2.setText(String.valueOf(x*x));
    }
}
usr_inp.html
<html>
```

ISBN: 978-93-91373-84-9

```
<body>
<applet code="usr_inp.class" width="400" height="300">
</applet>
</body>
</html>
```

### 9.13. Graphics Programming

- One of the most important feature of Java is its ability to draw graphics.
- We can write Java applets to draw lines, figures of different shapes, images and text in different fonts and styles.
- Every applet has its own area of the screen known as canvas, where it creates the display.

#### 9.13.1. The Graphics Class

- Java's Graphics class includes methods for drawing different types of shapes, from simple lines to polygons to text in a variety of fonts.
- To draw a shape on the screen we have to call one of the methods available in the Graphics class

| Method | Description |
|---|---|
| clearRect() | Erases a rectangular area of the canvas |
| copyArea() | Copies a rectangular area of the canvas to another area |
| drawArc() | Draws a hollow arc |
| drawLine() | Draws a straight line |
| drawOval() | Draws a hollow oval |
| drawPolygon() | Draws a hollow polygon |
| drawRect() | Draws a hollow rectangle |

ISBN: 978-93-91373-84-9

### 9.13.2. Lines and Rectangles

- The simplest shape we can draw with the Graphics class is a line.
- The drawLine() method takes two pair of coordinates (x1,y1) and (x2,y2) as arguments and draws a line between them.
- Example:
  - g.drawLine(10,10,50,50);

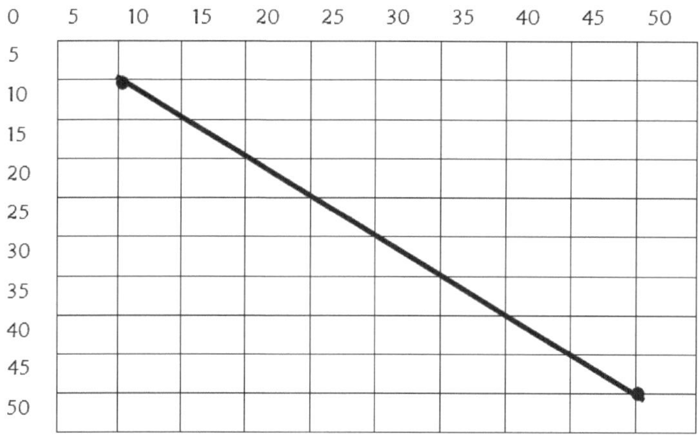


- The above statement in the coordinate system would draw a line as shown above.
- We can draw rectangle using the drawRect() method. This method takes four arguments.
- The first two represent the x and y coordinates of the top left corner of the rectangle, and the remaining two represents the width and the height of the rectangle.
- Example
  - g.drawRect(10,30,35,25);

ISBN: 978-93-91373-84-9

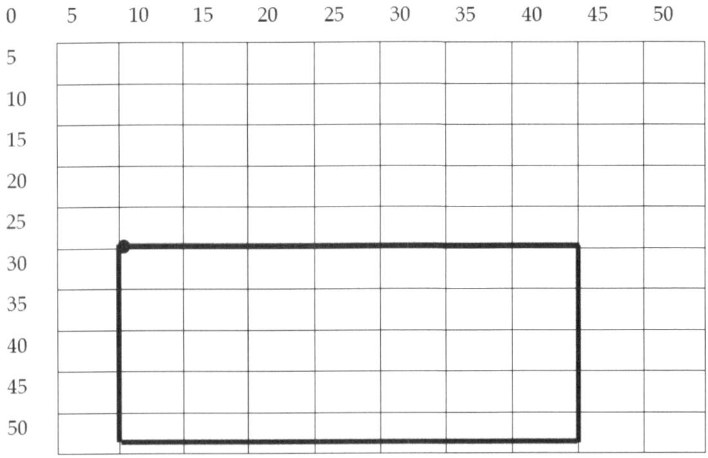


- The above statement in the coordinate system would draw a rectangle as shown above.
- We can draw a solid box by using the method fillRect() which also takes four arguments, corresponding to the starting point, width and height of the rectangle.
- We can also draw rounded rectangles using the methods drawRoundRect() and fillRoundRect().
- These two method take six arguments, corresponding to the starting point, width, height and the width and height of the angle of corners
- Example
    - g.drawRoundRect(10,100,80,59,10,10);
    - g.fillRoundRect(20,110,60,30,5,5);

### 9.13.3. Circles and Ellipses

- The graphics class does not have any method for circles or ellipses.
- The drawOval() method can be used to draw circles and ellipse.

ISBN: 978-93-91373-84-9

- Ovals are just like rectangles with overly rounded corners. The drawOval() method takes four arguments, the first two represent the top left corner of the imaginary rectangle and the other two represents the width and height of the oval itself.
- The fillOval() draws a solid oval
- Example
    - g.drawOval(20,20,60,40);

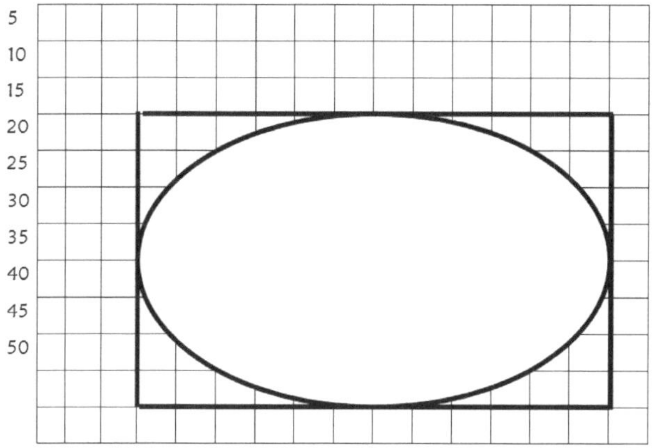


- We can set color of the solid oval using the method setColor()
- Example
    - g.setColor(Color.green);

### 9.13.4. Drawing Arcs

- An arc is a part of an oval. We can think of an oval as a series of arcs that are connected together in an orderly manner
- The drawArc() method takes six arguments. The first four are the same as the arguments of drawOval() method and the last two represent the starting angle of the arc and the number of degrees around the arc.

ISBN: 978-93-91373-84-9

- Java considers the 3 O'clock position as zero degree position and the degrees increase in anti-clockwise direction
- Example
    - g.drawArc(20,20,50,30,0,-180)

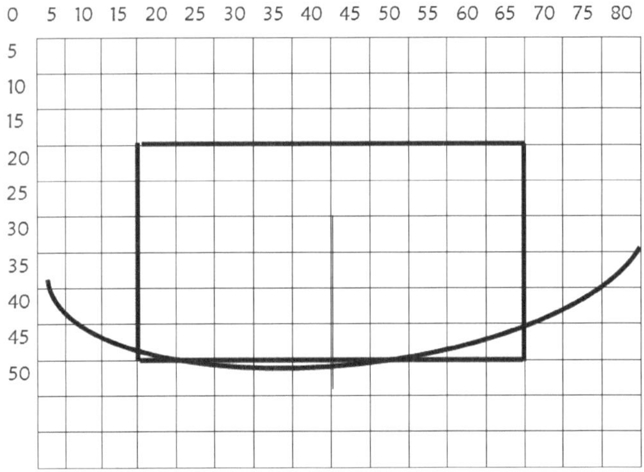


### 9.13.5. Drawing Polygons

- Polygons are shapes with many sides. A polygon may be considered a set of lines connected together
- The end of the first line is the beginning of the second line, the end of the second is the beginning of the third, and so on.
- We can draw polygon using the drawPolygon() method of the Graphics class.
- This method takes three arguments
    - An array of integers containing x coordinates
    - An array of integers containing y coordinates
    - An integer for total number of points
- Example
    - int p1[]={10, 70,80,10};

ISBN: 978-93-91373-84-9

- int p2[ ]={20,40,55,20};
- int n[]=p1.length;
- g.drawPolygon(p1,p2,n);

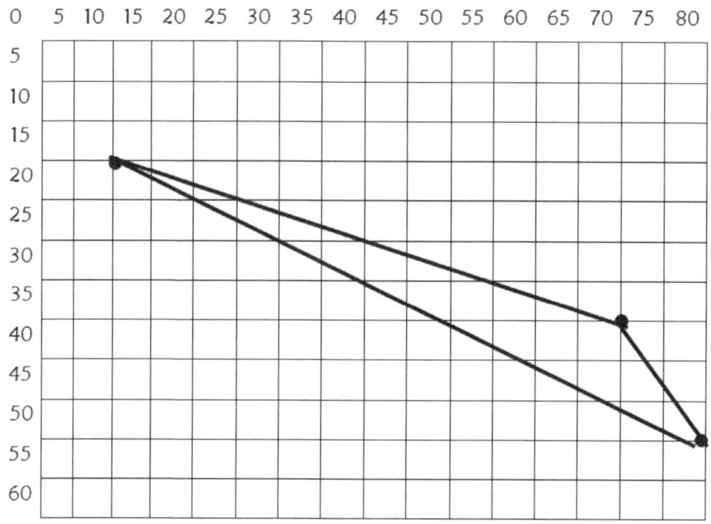


ISBN: 978-93-91373-84-9

# CHAPTER 10
# Managing Files

## 10. Managing Input/Output Files in Java

- We use variables and arrays to store data in the program. This approach poses the following problems
  - The data is lost either when the variable goes out of scope or when the program is terminated.
  - It is difficult to handle large volumes of data using variables and arrays
- This problem can be overcome by storing data on secondary storage devices such as floppy disks or hard disks. The data is stored in these devices using the concept of files.
- Data stored in files is called persistent data.
- A file is a collection of records placed in a particular area on the disk. A record is composed of several fields
- Storing and managing data using files is known as file processing which includes task such as creating files, updating files and manipulation of data.
- Reading and writing data in a file can be done at the level of bytes or characters or fields depending on the requirements of the application.
- The process of reading and writing objects is called object serialization.

## 10.1. Concept of Streams:

- A stream in Java is a path along which data flows. It has a source and a destination
- A stream can be defined as a sequence of data. Java streams are classified into two basic types

ISBN: 978-93-91373-84-9

- The InputStream is used to read data from a source and the OutputStream is used for writing data to a destination. InputStream and OutputStream are the basic stream classes in Java

### 10.1.1. Stream Classes:

- The java.io pacakage contains a large number of stream classes that provide capabilities for processing all types of data. These classes may be categorized into two groups on which they operate
  - Byte Stream classes that provide support for handling I/O operations on bytes
  - Character Stream classes that provide support for managing I/O operations on characters

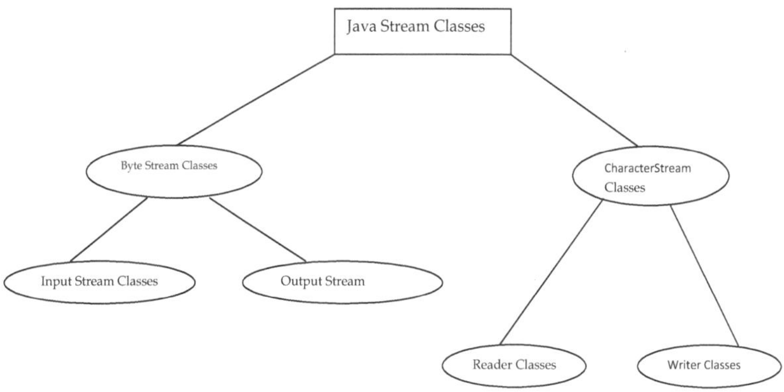


### 10.1.2. Byte Stream Classes

- This class is designed to provide functional features for creating and manipulating streams and files for reading and writing bytes.
- These streams are unidirectional and so they can transmit bytes in only one direction.
- Java provides two kinds of bytes stream classes
  - Input stream classes
  - Output stream classes

ISBN: 978-93-91373-84-9

#### 10.1.2.1.1. Input stream classes

- These classes are used to read 8-bit bytes. They include a super class known as Input Stream and a number of subclasses for supporting various input-related functions
- The Input Stream is an abstract class and so we cannot create instances of this class.
- The Input Stream class defines methods for performing input functions such as
  - Reading bytes
  - Closing streams
  - Marking positions in streams
  - Skipping ahead in a stream
  - Finding the number of bytes in a stream
- The methods in Input Stream class are
  - read() - reads a byte from the input stream
  - available() - gives number of bytes available in the input
  - skip(n) - skips n bytes from the input stream
  - close() - closes the input stream

ISBN: 978-93-91373-84-9 

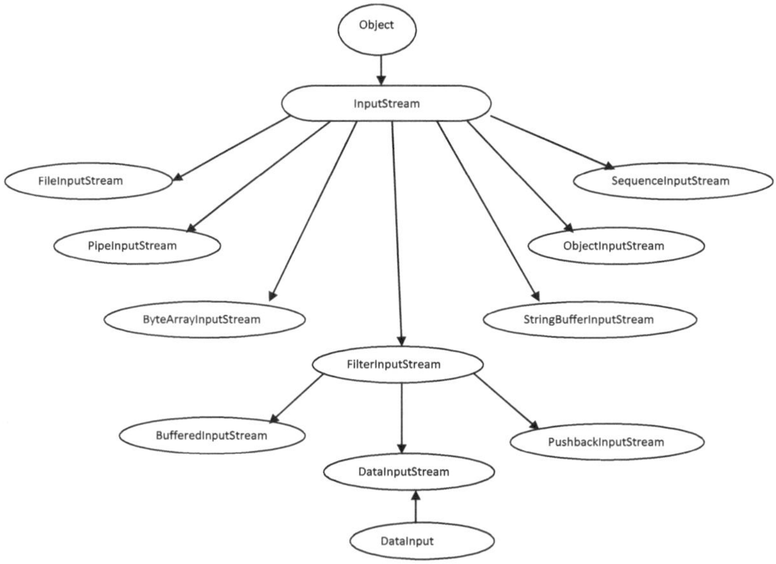


- The class DataInputSteam is a subclass of FilterInputStream and it implements the interface DataInput.
- The DataInput interface contains the following methods
  - readInt()
  - readLine()
  - readChar()

### 10.1.2.1.2. Output Stream classes

- These classes are derived from the base class OutputStream. This class is an abstract class and so it cannot be instantiated.
- The OutputStream includes methods that peform the following tasks
  - Writing bytes
  - Closing streams
  - Flushing streams

ISBN: 978-93-91373-84-9

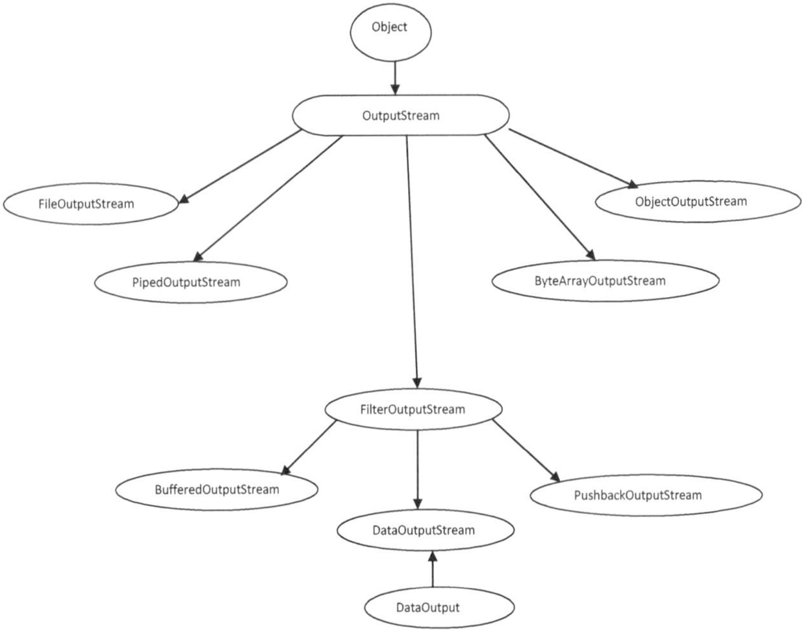


- The methods defined by OutputStream class are
  - write() - write bytes to the output stream
  - close() - closes the output stream
  - flush() - flushes the output stream
- The DataOutputStream implements the interface DataOutput. Some of the methods contained in DataOutput interface is
  - writeshort()
  - writeInt()
  - writeChar() etc

### 10.1.3. Character Stream Classes

- This class was not a part of the language when it was released in 1995.

ISBN: 978-93-91373-84-9

- Character streams can be used to read and write 16-bit Unicode characters.
- There are two kinds of character stream classes
  - Reader stream
  - Writer stream

### 10.1.3.1. Reader Stream Classes

- This class is designed to read characters from the files.
- Reader class is the base class for all other classes in this group.
- The reader class can perform all the functions implemented by the input stream classes

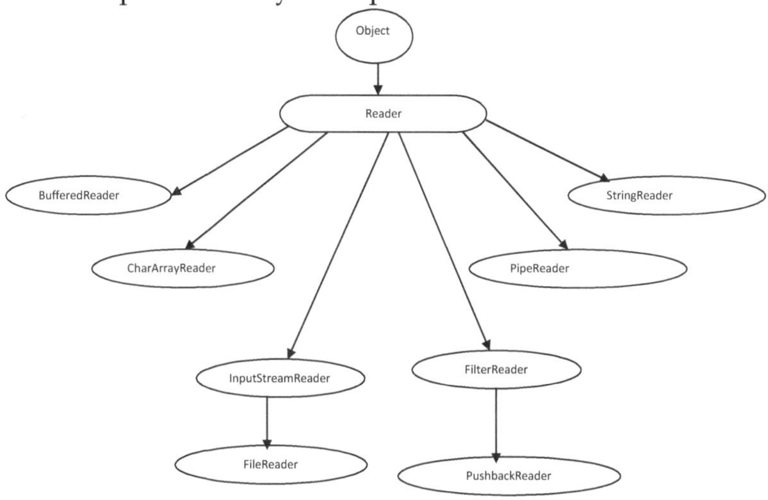


### 10.1.3.2. Writer Stream Classes

- The writer stream class is designed to perform all output operations on files.
- The writer stream classes are designed to write characters

ISBN: 978-93-91373-84-9

- The writer class is an abstract class which acts as a base class for all other writer stream classes.

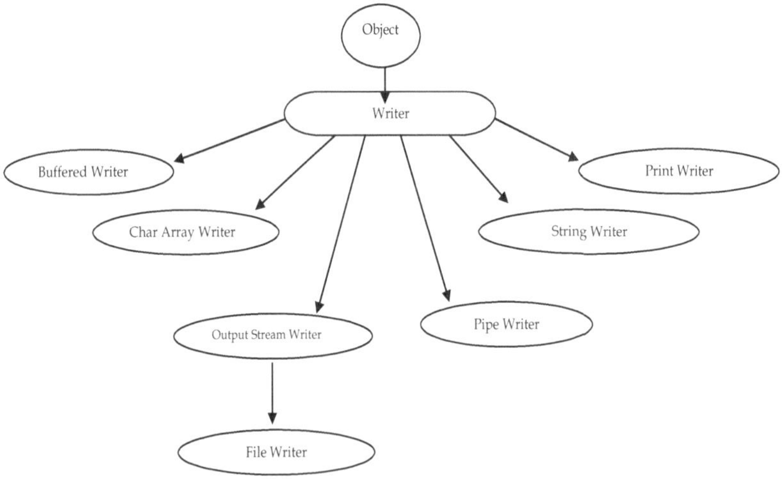


## 10.2. Other useful I/O classes

- The java.io package supports other classes for performing specialized functions
  - RandomAccessFile - this helps us to read and write bytes, text and Java datatypes to any location in a file
  - StreamTokenizer - is used for breaking up a stream of text from an input text file into meaningful pieces called tokens.

### 10.2.1. Using the File Class

- The java.io package includes a class known as File class that provides support for creating files and directories.
- This class contains several methods for supporting the operations such as
  - Creating a file

- Opening a file
- Closing a file
- Deleting a file
- Getting the name of a file
- Getting the size of a file
- Checking the existence of a file
- Renaming a file
- Checking whether the file is writable
- Checking whether the file is readable

## 10.3. Creating a File

- If we want to create and use a disk file we need to do the following
  - Suitable name for the file - which is helpful in identifying the file on the disk
  - Data type to be stored - helps to decide the file stream classes to be used for handling the data
  - Purpose (reading, writing or updating) - to know whether the file is created for readingonly, writing only or both
  - Method of creating the file -

**Reading / Writing Characters:**

- The classes Reader and Writer can handle characters
- Example: Program for copying the contents of one file to another

```
import java.io.*;
class co
{
        public static void main(String a[])
        {
                File inFile=new File("inp.dat");
```

```
        File outFile=new File("out.dat");
        FileReader ins=null;
        FileWriter ou=null;
            try
            {
            ins= new FileReader(inFile);
            ou=new FileWriter(outFile);
            int ch;

                while((ch=ins.read())!=-1)
                {
                    ou.write(ch);
                }
            }
            catch(IOException e)
            {
                System.out.println(e);
                System.exit(-1);
            }

            finally
            {
                try
                {
                ins.close();
                ou.close();
                }
                catch(IOException e)
                {       }
            }
        }
    }
```